DETECTION OF CHANGE
Event-Related Potential and fMRI Findings

DETECTION OF CHANGE
Event-Related Potential and fMRI Findings

edited by

John Polich, Ph.D.
The Scripps Research Institute, La Jolla, California

KLUWER ACADEMIC PUBLISHERS
Boston / Dordrecht / New York / London

Distributors for North, Central and South America:
Kluwer Academic Publishers
101 Philip Drive
Assinippi Park
Norwell, Massachusetts 02061 USA
Telephone (781) 871-6600
Fax (781) 681-9045
E-Mail: kluwer@wkap.com

Distributors for all other countries:
Kluwer Academic Publishers Group
Post Office Box 322
3300 AH Dordrecht, THE NETHERLANDS
Telephone 31 786 576 000
Fax 31 786 576 254
E-Mail: services@wkap.nl

QP
376
.5
. D48
2003

 Electronic Services < http://www.wkap.nl >

Library of Congress Cataloging-in-Publication Data

Title: DETECTION OF CHANGE: Event-Related Potential and fMRI Findings
Editor: John Polich
ISBN: 1-4020-7393-3

A C.I.P. Catalogue record for this book is available
from the Library of Congress.

Table of Contents

Contributors...vii

Introduction ix

MISMATCH NEGATIVITY

Chapter 1 1
AUDITORY ENVIRONMENT AND CHANGE DETECTION AS
INDEXED BY THE MISMATCH NEGATIVITY (MMN)
Anu Kujala and Risto Näätänen

Chapter 2 23
EVENT-RELATED BRAIN POTENTIAL INDICES OF
INVOLUNTARY ATTENTION TO AUDITORY STIMULUS
CHANGES
Kimmo Alho, Carles Escera, and Erich Schröger

Chapter 3 41
VISUAL MISMATCH NEGATIVITY
Dirk J. Heslenfeld

Chapter 4 61
CHANGE DETECTION IN COMPLEX AUDITORY
ENVIRONMENT: BEYOND THE ODDBALL PARADIGM
István Winkler

P3A AND P3B

Chapter 5 83
THEORETICAL OVERVIEW OF P3a AND P3b
John Polich

Chapter 6 99
    LATERAL AND ORBITAL PREFRONTAL CORTEX
    CONTRIBUTIONS TO ATTENTION
    Kaisa M. Hartikainen and Robert T. Knight

Chapter 7 117
    ERP AND fMRI CORRELATES OF TARGET AND NOVELTY
    PROCESSING
    Bertram Opitz

EEG, MEMORY, AND GAMMA

Chapter 8 133
    EEG AND ERP IMAGING OF BRAIN FUNCTION
    Alan Gevins, Michael E. Smith, and Linda K. McEvoy

Chapter 9 149
    EEG THETA, MEMORY, AND SLEEP
    Wolfgang Klimesch

Chapter 10 167
    GAMMA ACTIVITY IN THE HUMAN EEG
    Christoph S. Herrmann

Index 185

Contributors

KIMMO ALHO
Cognitive Brain Research Unit
Department of Psychology
University of Helsinki

CARLES ESCERA
Department of Psychiatry and Clinical Psychobiology
University of Barcelona

ALAN GEVINS
San Francisco Brain Research Institute &
SAM Technology

KAISA M. HARTIKAINEN
Department of Psychology and
Helen Wills Neuroscience Institute
University of California, Berkeley

CHRISTOPH S. HERRMANN
Max-Planck-Institute of Cognitive Neuroscience
Leipzig

DIRK J. HESLENFELD
Department of Psychology
Free University

WOLFGANG KLIMESCH
Department of Physiological Psychology
University of Salzburg

ROBERT T. KNIGHT
Department of Psychology
Helen Wills Neuroscience Institute
University of California, Berkeley

ANU KUJALA
Cognitive Brain Research Unit
Department of Psychology
University of Helsinki

LINDA K. McEVOY
San Francisco Brain Research Institute &
SAM Technology

RISTO NÄÄTÄNEN
Cognitive Brain Research Unit
Department of Psychology
University of Helsinki

BERTRAM OPITZ
Experimental Neuropsychology Unit
Saarland University

JOHN POLICH
Department of Neuropharmacology
The Scripps Research Institute

ERICH SCHRÖGER
Institute for General Psychology
University of Leipzig

MICHAEL E. SMITH
San Francisco Brain Research Institute &
SAM Technology

ISTVÁN WINKLER
Institute of Psychology
Hungarian Academy of Sciences
Cognitive Brain Research Unit, Department of Psychology
University of Helsinki

INTRODUCTION

1.   DETECTION OF CHANGE

As sensory stimuli are experienced, adaptive neural mechanisms extract information from these events. However, the processes underlying this capability are not yet well understood and continue to inspire research efforts. Measures of brain activity when a change in stimulation occurs can be assessed with electroencephalography (EEG), event-related brain potential (ERP), and functional magnetic resonance imaging (fMRI) techniques. The chapters in this book are snapshots of the recent progress made with these methods.

The central theme is the detection of change when stimulus parameters are well controlled. The main questions are: Where and how does neural change detection occur? Are similar processes elicited across modalities? How do these events contribute to cognition? Leading experts have reviewed these issues, with background material integrated into each chapter. Topics include analysis of mismatch negativity, P3a/P3b theory and sources, human lesion studies, how EEG reflects cognition, and stimulus binding. These areas serve as the backdrop for discussions of stimulus modality ERP effects, the conjoint use of fMRI methods, and neuroelectric models of attention, perception, and memory.

2.   ORGANIZATION AND CONTENTS

The text covers the gamut of experimental studies using stimulus change paradigms, with clinical data augmenting the utility of the methods. The book's chapters are organized around the major topics of MMN, P300, and EEG oscillations to provide a spectrum on how modern neuroimaging methods can measure stimulus change processing. The authors constructed the chapters as they deemed appropriate but were encouraged to write for a broad audience by reviewing results in their theoretical context. This goal was very well met, so that the contents are fresh and the literature distillations helpful and informative.

The first section on MMN provides a very assessable précis of this huge ERP research area. Kujala and Näätänen lead off with a synopsis of the field that sets the stage for the subsequent chapters. The historical developments of the MMN, its theory, and clinical applications are clearly limned. Alho, Escera, and Schröger then examine the relationships among MMN, P3a, and reorienting negativity produced by auditory/visual stimulus interactions. The measurement precision and integrative results are illuminative and important. Heslenfeld describes the background for and the data from a series of innovative studies that appear to elicit the elusive visual MMN. The results are exciting and the implications for future MMN work intriguing. Winkler thoughtfully tackles the fundamental assumptions underlying MMN by going "beyond the oddball paradigm." How a deviant stimulus can be defined is discussed with compelling scholarly force in this provocative chapter.

The second section extends these topics by dissecting the P300 into its constituent P3a and P3b subcomponents. Polich presents an overview of P300 theory and outlines how the P3a and P3b may interact. Stimulus novelty *per se* is not required for P3a generation under appropriate task conditions, so the psychological origins of this potential and the P3b can be reasonably inferred. Hartikainen and Knight cogently review ERP data from neurologically lesioned patients. The findings delineate how different brain structures contribute to P3a and P3b and are of keen theoretical interest. Opitz reports on P300 studies that integrate ERP and fMRI methods in normal subjects. The data from both approaches are constrained by current source density analysis to help isolate P3a and P3b neural loci in a technically rigorous fashion

The third section focuses on EEG oscillations. Gevins, Smith, and McEvoy succinctly summarize EEG methods by highlighting how advanced techniques can magnify the sensitivity of this brain measure. The findings forcefully demonstrate that increased resolution and sophisticated analysis clearly abet cognitive neuroscience. Klimesch provides an informative review of the relationship between EEG and memory processes. Event-related desynchronization (ERS) data from sophisticated designs appear to reflect the genesis of memory formation. Hermann describes the technical basis of gamma activity and how it may underlie stimulus binding. The illustrative studies strongly support the excitement of the "gamma bandits" that the origins of perceptual consciousness can be measured.

3.   FINAL COMMENTS

As this summary suggests, the book's chapters encapsulate the recent findings on how electric and magnetic measures reflect detection of stimulus change. Putting this project together has been immensely stimulating and rewarding. I sincerely and very much thank all of the authors for their contributions and patient collegial support. The superb technical skills of Angela Caires and Nancy Callahan are gratefully acknowledged. I also thank Floyd Bloom for helping to make all this possible.

John Polich
La Jolla, California
September, 2002

Chapter 1

AUDITORY ENVIRONMENT AND CHANGE DETECTION AS INDEXED BY THE MISMATCH NEGATIVITY (MMN)

ANU KUJALA AND RISTO NÄÄTÄNEN
Cognitive Brain Research Unit, Department of Psychology, University of Helsinki, Finland

Mismatch negativity (MMN) is an automatic event-related brain potential (ERP) that reflects a change in auditory stimulation and provides a unique measure of central sound representation. The electrically registered MMN and its magnetic equivalent MMNm are elicited by a discriminable change in any repetitive aspect of auditory stimuli even in the absence of attention (Näätänen et al., 1978; Näätänen, 1992). Moreover, the brain mechanisms generating the MMN response initiate an attention switch to sound change, and thus cause its conscious perception. Hence, MMN can be used to probe the emergence and accuracy of the cortical representations for present and past sound events. Furthermore, the MMN can be used as a means to assess deficits in central auditory processing for various clinical conditions.

The auditory environment is almost continuously changing. A change can take place within the stream of sounds reaching the auditory system (e.g., in a speech stream or just in the background noise originating from the street). A change can also be an appearance of a new or unexpected sound (e.g., a new warning signal or a cough in the middle of chorus melody), a modulation in an ongoing familiar sound (e.g., a rise in speech voice), or even an omission of a repetitive sound in a sound stream. The common factor for all these situations is that before the change, a somewhat stable auditory environment existed. For a change to be detected, the context to which it is compared must be represented, even though this context usually is also changing, at least with respect to some of its features. How, then, does the brain form the auditory context and how is a change in it detected and distinguished as a potentially relevant event?

1.   MMN: THE BRAIN'S AUTOMATIC RESPONSE TO CHANGES IN AUDITORY STIMULATION

The MMN and its magnetic equivalent MMNm are elicited by any discriminable change in some repetitive aspect of auditory stimulation. In a traditional MMN paradigm, infrequent (deviant) stimuli occasionally replacing the repeating (standard) stimuli elicit an MMN, which peaks at 100-200 milliseconds from change onset. This stimulus change produces a negative deflection in the ERP to deviant stimuli relative to the standard-stimulus waveform. The MMN is usually separated from other ERP components that the standard and deviant stimuli elicit by subtracting the standard ERP waveform from that to the deviant stimuli. Thus, the remaining difference waveform is related only to the stimulus change, although some contribution of changes in the exogenous components can in certain cases occur.

Figure 1 illustrates how MMN amplitude and latency depend on the magnitude of the stimulus change. MMN amplitude is enhanced and latency shortened as the difference between the deviant and standard stimuli is increased. Furthermore, MMN is elicited even in the absence of attention, as when the subject is reading, watching a silent video, or performing a visual task, or even when a patient is in a coma state (Kane et al., 1993). Attention can, however, have some influence on the MMN amplitude, but its withdrawal does not abolish the MMN (Trejo et al., 1995; Woldorff et al., 1991).

It is important to emphasize that stimulus change causes the MMN, with the infrequent sounds alone eliciting no MMN (Korzyukov et al., 1999; Kraus et al., 1993; Näätänen et al., 1989a). Accordingly, MMN can be produced by stimuli occasionally occurring too early in a stimulus sequence (Ford & Hillyard, 1981; Hari et al., 1989; Näätänen et al., 1993a; Nordby et al., 1988) or when omitted from a rapidly presented stimulus train (Yabe et al., 1997). The MMN is therefore a response that reflects change detection in the automatic comparison of the present stimulus with the sensory-memory representation of the previous stimuli. Hence, MMN is not produced by deviant stimuli activating new afferent sensory elements unrelated to those activated by the standard stimuli. Furthermore, electrical (Giard et al., 1995), magnetic (for a review, see, e.g., Alho, 1995), and intracranial (Kropotov et al., 1995, 2000) recordings have shown that this change-detection process originates in the auditory cortices, with some evidence obtained for an additional right-hemispheric frontal MMN generator (Alho et al., 1994; Giard et al., 1990; Rinne et al., 2000).

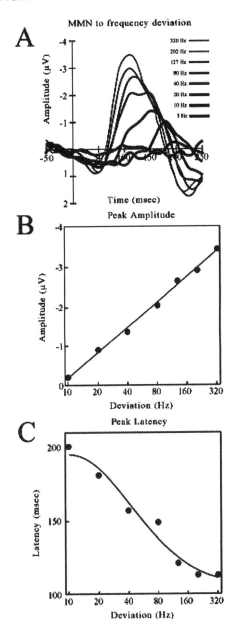

Figure 1. (a) MMN to frequency deviation: grand-average difference waves obtained by subtracting ERPs to 1000 Hz standard tones from those to deviant tones with higher frequencies (see legend). Each deviant stimulus occurred among standard tones in separate stimulus blocks at a probability of 0.05 (data from 10 subjects). MMN amplitude increases and latency decreases with increasing frequency deviation. (b) MMN peak amplitude increases with increases in the magnitude of frequency deviation. (c) MMN peak latency decreases as the magnitude of frequency deviation increases (after Tiitinen et al., 1994).

2.   CENTRAL SOUND REPRESENTATIONS AS INDEXED BY MMN

The MMN can be elicited by a change in sound frequency (Sams et al., 1985; Tiitinen et al., 1994), intensity (Lounasmaa et al., 1989; Näätänen et al., 1987a), or spatial locus of origin (Paavilainen et al., 1989; Schröger & Wolff, 1996). The MMN is also elicited by changes in the frequency components of complex sounds such as phonemes (Aaltonen et al., 1987; 1994; Aulanko et al., 1993; Kraus et al., 1995; Näätänen et al., 1997) or other spectrally complex sounds (Alho et al., 1996; Winkler et al., 1998), such as chords (Tervaniemi et al., 1999). MMN is also elicited by changes in the temporal features of sound stimulation, such as duration (Kaukoranta et al., 1989; Näätänen et al., 1989b), rise time (Lyytinen et al., 1992), the temporal structure of sound patterns (Alho et al., 1993; 1996; Näätänen et al., 1993b; Schröger, 1994; Tervaniemi et al., 1997; Winkler & Schröger, 1995), or shortening of the time interval between successive stimuli. Even violations of abstract relationships between the elements of auditory stimulation can evoke the change-detection process, as when an infrequent tone repetition occurs within a sequence of tones with a continuously descending pitch (Tervaniemi et al., 1994). Furthermore, MMN studies have shown that the traces of sound representations contain feature-integrated information (Gomes et al., 1997; Sussman et al., 1998; Takegata et al., 1999). In these studies, subjects were presented with two types of standard tones differing from each other (e.g., in frequency and intensity). The deviants possessed one feature of each standard and therefore formed a deviant conjunction of the frequent levels of the two attributes to elicit an MMN. Taken together, these findings suggest that the cortical traces reflected by the MMN contain integrated spectral, temporal, and even abstract information about sound events.

3.   ACCURACY OF THE NEURAL REPRESENTATIONS FOR SOUNDS

There is a narrow range of any standard stimulus feature within which a deviant stimulus elicits no MMN (Näätänen & Alho, 1995, 1997). Figure 2 illustrates this phenomenon, which is termed the representational width (Rw). For a typical young adult, the MMN Rw is about 0.5-2.0% for 1000 Hz standard tones, so that deviant tones between the 995-1005 Hz (Rw=0.5%) frequency range usually do not activate the change-detection process. That is, the deviants elicit a small, nonsignificant MMN and are not usually consciously discriminated. The individual sharpness (informational specificity) of the sound representations can be, at least theoretically, defined with the MMN: the narrower the Rw, the sharper and more stimulus-specific is the sound representation in the brain.

MMN as a Function of Frequency Change

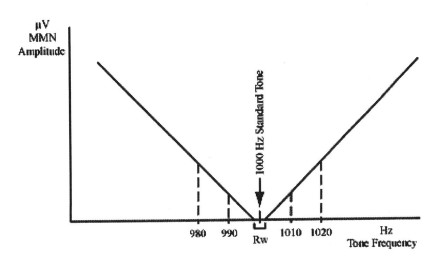

Figure 2. Schematic illustration of representational width (Rw) along a sensory dimension. Rw is the range around the standard-stimulus level in a given sensory feature, within which a deviant stimulus elicits no MMN (after Näätänen & Alho, 1997).

Figure 3 illustrates the relationship between behavioral discrimination ability and the MMN (Lang et al., 1990). This study employed three groups of 17 year old high-school students according to their behavioral pitch-discrimination performance ('good', 'moderate', 'poor'), and recorded the MMN to frequency changes of different magnitude in a separate session. In the good-performer group, a frequency deviation of 19 Hz was enough to elicit MMN, whereas a deviation from 50 to 100 Hz was needed in the poor-performance group for MMN elicitation; the moderate performers fell between these two extreme groups. In subsequent studies, MMN amplitude was shown to correlate with the behavioral discrimination of rhythmic sound patterns (Tervaniemi et al., 1997) and within-category examples of a vowel (Aaltonen et al., 1994). Corresponding results were obtained in studies demonstrating the emergence of the MMN when subjects learned to discriminate a change in a complex spectro-temporal tone pattern (Näätänen et al., 1993b) or when subjects learned to discriminate different variants of a consonant-vowel (/da/) syllable (Kraus et al., 1995). Thus, MMN is sensitive to individual auditory differences, as well as to the emergence of discrimination capability.

Discrimination Ability

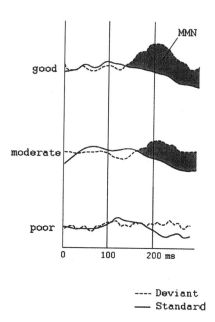

Figure 3. Grand-average ERPs for subjects who were 'good', 'moderate', or 'poor' in behavioral pitch-discrimination of infrequent deviant tones from standard tones in a difficult pitch discrimination task. MMN amplitude and latency (shaded areas) for the frequency deviation differs among the groups (after Lang et al., 1990).

4.   EMERGENCE OF SOUND REPRESENTATIONS

As suggested by the loudness summation of tones (Scharf & Houtsma, 1986) and the test-to-masking stimulus interval in backward-masking studies (Hawkins & Presson, 1986), the temporal window of integration in auditory perception has a duration of 150-200 milliseconds. Figure 4 schematically illustrates the formation of the cortical trace underlying the sound perception. The MMN has also been used in determining the time needed for the emergence of the central sound representation. Winkler and Näätänen (1992) presented tones that were followed by a masker tone presented afterwards at varying time intervals. MMN was elicited when the silent time interval between the standards and deviants and the masker was 150 milliseconds or longer but not at the shorter intervals. The masker tone presumably prevented trace formation, since with time intervals shorter than 150 milliseconds subjects were unable to behaviorally discriminate the deviants. Subsequent MMN studies (Schröger, 1997; Yabe et al., 1997, 1998; Sussman et al., 1999) have verified that the temporal window of integration, as estimated from MMN data, has a duration corresponding to that suggested by the behavioral studies. These findings suggest that this integration time is necessary for acoustical features to form a unitary auditory event, rather than being represented as static features for each time point (Näätänen, 1992).

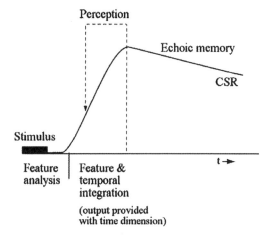

Figure 4. A schematic illustration of the emergence and decay of sound central auditory representation (CSR). First, sound attributes are rapidly mapped on the respective separate feature analyzers whose outputs are subsequently mapped on the neurophysiological mechanisms of sensory memory so that the basis for unitary sound perception emerges through feature and temporal integration. The emerging sound representation has a time dimension, as sounds are represented as events in time rather than as individual static features. The emergence of this sound representation provides the specific information contents for the sound percept (after Näätänen & Winkler, 1999).

5. SHORT-TERM AND LONG-TERM MEMORY TRACES FOR SOUND REPRESENTATIONS

Modeling of the auditory environment in the form of sound representations is based on the memory traces of short-term sensory memory. The MMN is thought to reflect these traces and thereby provides an indicator for their development and possible neuroanatomical locations. The MMN is elicited when a discrepancy is found between the input from a deviant stimulus and the trace of the standard stimulus. This outcome, however, requires that the standard-stimulus trace has not decayed and, therefore, the standard sounds must be repeatedly presented at relatively short intervals. The duration of these short-term traces is estimated to be of the order of 10 seconds (Cowan et al., 1993; Sams et al., 1993; see also Näätänen et al., 1987b). MMN data also suggest that sensory memory can maintain more than one sound representation in parallel, such that at least two (Sams et al., 1984; Winkler et al., 1996a) or even several (Gomes et al., 1997; Ritter et al., 1995) memory traces can be simultaneously active.

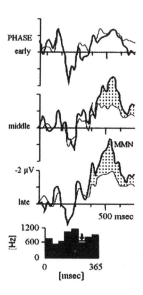

Figure 5. Grand-average ERPs from Cz to standard serial sound patterns (thin lines) and to deviant patterns (thick lines) occurring randomly with a probability of 0.1. The standard and deviant sound patterns consisted of 8 consecutive tones of different frequencies and are illustrated at the bottom of the figure. In the deviant patterns, the frequency of the sixth tone (indicated by the arrow) was higher than in the standard patterns. ERPs were recorded during the early, middle, and late phases of a session in which sound patterns (1,200 in each phase) were presented to subjects who were reading. The performance of the subjects belonging to this group improved during the session in a sound-pattern discrimination test applied after each phase of the session. MMN (shaded area) first emerged and then increased in amplitude during the session (after Näätänen et al., 1993b).

One presentation of the standard stimulus seems to be enough to reactivate the standard-stimulus trace if it has decayed to the extent that deviants no longer elicit MMN (Cowan et al., 1993; Winkler et al., 1996a). This finding indicates that the standard-stimulus trace formed in auditory short-term sensory memory has coalesced into a durable form of sensory memory (Cowan et al., 1993). Figure 5 illustrates this effect, with a longer-term learning effect for MMN generation demonstrated by using complex spectro-temporal sound patterns and a minor deviation in one tone component (Näätänen et al., 1993b). Passive (subjects reading) MMN-recording blocks were alternated with active (deviant detection) sound-discrimination blocks, which produced MMN emergence as subjects started behaviorally to discriminate deviants from standards. Thus, longer-term memory traces are slowly formed with attentive training, as the passive exposure *per se* was not sufficient to cause MMN emergence.

6.   LANGUAGE-SPECIFIC SPEECH-SOUND TRACES

Long-term auditory traces are presumably crucial for speech perception and serve as recognition patterns for speech-sound segments, in which some invariant relationships among the acoustic elements rather than sensory information *per se* is encoded (Näätänen, 2001). Figure 6 summarizes MMN findings that suggest the existence of permanent memory traces for native-language vowels (Näätänen et al., 1997). In this study, Estonian and Finnish subjects were presented with phoneme contrasts. When the standard phoneme /e/ (shared by both languages) was replaced by a phoneme of the Estonian language /õ/, MMN was larger in Estonian than Finnish subjects, whereas it did not differ between the two groups when the deviant stimulus was a phoneme belonging to both languages (/ö/ or /o/). Corroborating results have been obtained with the French (Dehaene-Lambertz, 1997) and Hungarian (Winkler et al., 1999b) languages. In addition to phonemes and phonological units (Phillips et al., 2000; Dehaene-Lambertz et al., 2000), memory traces for words are also reflected by the MMN (Pulvermüller et al., 2001), since larger-amplitude MMN was found in native Finnish speakers to a syllable change producing a two-syllable Finnish word than a non-word, with the effect being absent in subjects who did not understand Finnish.

Learning a new language can be monitored with the MMN, which appears to reflect development of the phonemic cortical memory representations of the new language. MMN has been observed to a contrast between two Finnish vowels in adult Hungarians who were fluent Finnish speakers, but not in Hungarians who had no experience of Finnish (Winkler et al., 1999a). Hence, in those Hungarians fluent in Finnish, the cortical memory representations of the Finnish vowels had developed as they learned this language. In children, speech-sound traces for the mother tongue, as indexed by the enhanced MMN elicitation when the deviant stimulus is a phoneme of this language, are formed between 6-12 months (Cheour et al., 1998a) or even earlier in infancy (Dehaene-Lambertz & Baillet, 1996)—well before the speech production properly starts. Taken together, these results demonstrate the sensitivity of MMN to language-related stimulus processing.

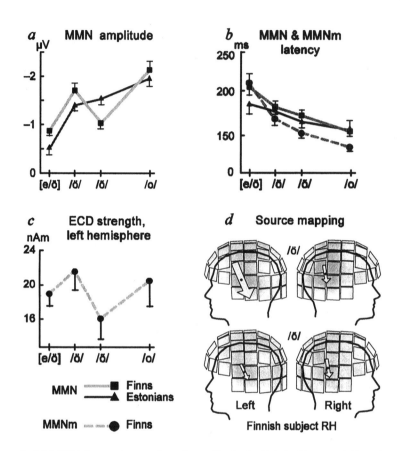

Figure 6. (a) MMN (mean + s.e.m.) peak amplitude (at Fz) in Finns and Estonians as a function of the deviant stimulus, arranged in the order of increasing F2 difference from the standard stimulus. (b) MMN (at Fz, solid lines) and MMNm (left hemisphere; broken line) peak latencies as a function of the deviant stimulus for Finns and Estonians. (c) Strength of the equivalent current dipole (ECD) modeling the left auditory-cortex MMN for the different deviant stimuli (n=9). (d) Left- and right- hemisphere MMNms of one typical Finnish subject for deviants /ö/ and /õ/ presented in contour (spacing 2 fT/cm) maps of the magnetic field-gradient amplitude at the MMNm peak latency. The squares indicate the arrangement of the magnetic sensors. The arrows represent ECDs indicating activity in the auditory cortex; the black dots in these arrows show the centers of gravity of the MMNm. Note that the prototype /ö/ elicits a much larger MMNm in the left than in the right hemisphere, whereas non-prototype /õ/ responses in both hemispheres are small and quite similar in amplitude (after Näätänen et al., 1997).

7.   NEURAL LOCI OF SHORT- AND LONG-TERM AUDITORY MEMORY TRACES

Several MMN studies suggest specialization within and between the hemispheres already at a very early and automatic level in processing different types of auditory information. The neural locations of the traces for the different properties of the same sound produce differences in the MMN sources for changes in these attributes (Giard et al., 1995; Levänen et al., 1993; Paavilainen et al., 1991). Magnetic recordings of the MMNm elicited by a change in simple vs. complex sounds also suggest that the sensory-memory representations of these sounds are located in different parts of the auditory cortex (Alho et al., 1996). In addition, recent MEG results indicate that the traces for phonemes and chords matched in frequency differ in location in both hemispheres (Tervaniemi et al., 1999). This intra-hemispheric specialization has also been observed in patients with left-hemisphere lesions (Aaltonen et al., 1993): when the lesion was located in the posterior areas, no MMN to a phoneme change was elicited, whereas the frequency change of a simple tone elicited an MMN. In contrast, patients with anterior lesions showed an MMN to both phoneme and tone-frequency changes. Consistent with these findings, recent studies have indirectly located the traces of native-language vowels to the left hemisphere, in or near Wernicke's area. For example, in Finnish subjects, the left auditory cortex dominated the MMNm elicited by the native-language deviant vowels, whereas similar but smaller MMNms were generated in the auditory cortices of both hemispheres when the deviant stimulus was a vowel of the Estonian but not Finnish language (Näätänen et al., 1997). Additional evidence for left-hemispheric lateralization of vowel memory traces also has been reported by studies using different methodologies (ERPs, Rinne et al., 1999; MEG, Gootjes et al., 1999; positron-emission tomography or PET, Tervaniemi et al., 2000). Finally, the left hemisphere also dominates the processing of longer phonetic units: a stronger MMNm was elicited by consonant vowel-syllable changes in the left than right hemisphere (Alho et al., 1998; Shtyrov et al., 1998; 2000), with complimentary fMRI data (Celsis et al., 1999) showing a stronger left-hemispheric than right-hemispheric activation to CV-syllable changes.

These results imply that the acoustic features of the speech sounds are represented by memory traces formed in both hemispheres, whereas phonetic information (i.e., about the phonetic invariance; Aulanko et al., 1993) is represented in the left hemisphere. Consistent with this assertion, MMN elicited by the syllable /da/ was larger over the left than right hemisphere when the change signalled a phonetic change, in contrast to similar MMNs over the left and right hemispheres when the same syllable signalled a pitch change (Sharma & Kraus, 1995). Thus, the different features of the speech sounds, such as the linguistic versus prosodic features of speech, might be represented by separate traces. Consequently, the cortical traces probed by the MMN are excellent candidates for acting as recognition templates in speech perception (Näätänen, 2001).

8.   MMN AND ATTENTIVE CHANGE-DETECTION

The formation of short-term sound representations enables neural maintenance of auditory information over a period of several seconds, thereby creating the recent auditory past. This presence of the immediate auditory history is often described as echoic memory (Cowan, 1988) and is needed for recognizing sound events from the continuously varying acoustic signal. The auditory context modeled in the sound representations (Winkler et al., 1996b) provides the reference information against which any change is detected—an event that usually results in an attention switch to this change. Several studies support the idea that MMN elicitation is related directly to involuntary attention switching. For example, Schröger (1996) presented subjects with irrelevant auditory stimulation while they were performing an auditory primary task. When a minor frequency change, eliciting an MMN, preceded the to-be-detected target stimuli, reaction time was prolonged and hit rate decreased, indicating an attention shift to the irrelevant frequency change. Similar results were obtained with visual primary tasks (Alho et al., 1997; Escera et al., 1998, 2000). Consistent with these observations are findings that implicate the contribution of the frontal cortex, which is known to have a central role in the control of the direction of attention to MMN generation (for a review, see Alho, 1995). The frontal MMN generator is usually stronger on the right than left hemisphere (Giard et al., 1990). Interestingly, the time course of this frontal activation is slightly delayed relative to the auditory-cortex activation (Rinne et al., 2000). Thus, the auditory-cortex processes underlying automatic change detection trigger the frontal mechanisms of involuntary attention switch (Näätänen, 1990).

However, a sound change does not cause attention switch each time the change occurs, so that the stimulus change can remain consciously unperceived. This outcome occurs for several reasons: First, it may be that the deviant sound does not differ enough from that represented by the sensory-memory trace, either because the difference is very small or the neuronal trace is informationally too diffuse. That is, its representational width for the auditory attribute involved is too large in relation to the magnitude of the change (Näätänen & Alho, 1997). Second, the memory trace underlying the previous sounds may already have decayed, so that no automatic comparison process can take place. Third, when attention at the moment of change is intensively focused elsewhere, then the threshold for attention switch is presumably elevated (Lyytinen et al., 1992; Näätänen, 1991, 1992). Fourth, the excitability of the frontal MMN generator may be temporally decreased, for example, by alcohol (Jääskeläinen et al., 1996). The system appears to be finally tuned so as to engage attentional switching and produce MMN under conditions that ensure sufficient stimulus processing has occurred.

9. MMN AS A CLINICAL INDEX FOR CENTRAL AUDITORY PROCESSING

As the MMN process reflects the central sound representation underlying both sound perception and sensory memory, this ERP has been used to assess the specific nature and degree of auditory processing disorders as well as a probe for ascertaining the general state of the brain. The MMN can be recorded from newborns (Alho et al., 1990) and even from pre-term newborns (Cheour-Luhtanen et al., 1996) to provide a measure of central auditory processing for evaluation of children who are diagnosed or are at-risk for auditory processing deficits. For example, children with cleft-palate or other genetic disorders (Cheour et al., 1998b), infants with a risk for dyslexia (Leppänen et al., 1997), or children suffering from dysphasia (Korpilahti & Lang, 1994), show abnormalities in MMN. If auditory processing problems can be detected early in infancy, more time is obtained and therefore increased efficiency of rehabilitation in these children may be possible. The early identification of children with or at-risk for language problems is also an important application of MMN, as the first 2-3 years of life are critical for language development. For example, comparative data of auditory sensory-memory system development, with respect to language development are now available for normal infants (for a review, see Kraus & Cheour, 2000).

Recent MMN studies have indicated that the deficits of auditory processing underlying speech disorders may have a more general nature. In adult developmental dyslexics, MMN to a shortening of a tone interval in the midst of a 4-tone pattern was absent, whereas a well-developed MMN was elicited in controls—findings that implicate problems in the temporal integration of auditory information in dyslexics (Kujala et al., 2000). Similar results were obtained in left-hemisphere stroke patients, as the MMN to and the behavioral discrimination for duration decrements of harmonic tones was deteriorated (Ilvonen et al., 2001). The improvement of sound discrimination in cochlear-implant patients, as indexed by an increasing MMN amplitude after the implantation, further supports the important relation between the cortical discrimination accuracy and speech-perception ability (Ponton et al., 2000). Thus, MMN can facilitate the identification of deficits in the very basic cortical mechanisms of sound processing that appear to underlie speech-sound processing.

MMN also can reflect general brain functioning in aging, Alzheimer's disease, and Parkinson's disease (for a review, see Pekkonen, 2000), as MMN sensory-memory traces decay at an accelerated rate. MMN has been used to assay frontal-lobe damage (Alho et al., 1994), closed-head injuries (Kaipio et al., 2000), and the right-hemispheric damage causing the neglect syndrome (Deouell et al., 2000). As noted above, MMN can be used to predict the awakening of coma patients (Kane et al., 1993, 1996; Fischer et al., 2000; Morlet et al., 2000). In addition, MMN has proven useful in assessment of psychiatric conditions, such as schizophrenia (Shelley et al., 1991), depression (Ogura et al., 1993), and alcoholism (Ahveninen et al., 2000; Polo et al., 1999), wherein central auditory processing appears to be compromised. More generally, MMN reflects the temporary fluctuations in the general brain function caused by alcohol ingestion that attenuates MMN amplitude (Jääskeläinen et al., 1998), and specifically its frontal subcomponent (Jääskeläinen et al., 1996).

10.   CONCLUSIONS

The brain's automatic sound-change detection mechanism is indexed by the MMN, which is based on the sound-event representations carried by the auditory cortical memory traces. MMN is presumably elicited when a trace is formed by a deviant sound event in the sensory-memory system of the auditory cortex where an active sensory-memory trace for the repetitive aspects of the preceding auditory stimulation already exists (Näätänen, 1985). Hence, those sensory-memory neurons activated by the deviant sound that were not already involved in representing the standard stimulus (but were released from tonic inhibition by the emergence and presence of this representation), likely contribute to MMN generation (Näätänen, 1990). This comparison process is automatic and requires no attention directed to the sounds, although it usually results in an attention switch to the eliciting sound change.

MMN studies have demonstrated that acoustic features are represented in the sensory-memory system as a unitary sound event with integrated feature and temporal information (Näätänen & Winkler, 1999). Even the abstract regularities and rules of the constantly changing auditory environment are reflected by these cortical representations, which appear to correspond to the conscious perception of the sound event. This relation of the neural traces involved in MMN elicitation to conscious perception is confirmed by the generally good correspondence of MMN amplitude and latency with the behavioral discrimination performance (Lang et al., 1990; Tiitinen et al.,

1994; Winkler et al., 1997). Further, the formation phase of the representation can be related to sound perception, with the slowly decaying phase lasting for the few seconds underlying the sensory memory of the sound event (Näätänen & Winkler, 1999). Thus, the sound-event representations formed and stored by the cortical traces underlying MMN elicitation serve as a context for change detection and contribute to the conscious percept of the change.

Magnetic and intracranial recordings have confirmed that the main generators of the MMN response are located in the auditory cortices (Kropotov, 2000; reviewed in Alho, 1995), with the frontal generator related to the attention switch to the eliciting sound change (Giard et al., 1990; Rinne et al., 2000). Moreover, the neural substrates involved in MMN elicitation are spatially distributed according to the physical or abstract features of the sound events. These findings add to the growing evidence for specialization within and between the auditory cortices in representing sound events that differ in nature, as demonstrated by the differences in the strengths and locations/orientations of the MMN responses to different sound changes.

The individual variation in the behavioral sound-discrimination ability is directly associated with the individual variation in MMN emergence, suggesting appreciable individual variation in the accuracy of the cortical sound representations. Differences in accuracy presumably depend upon the receptive-field width of the afferent neurons feeding information to the sensory-memory system (Näätänen & Alho, 1997). Besides this normal inter-individual variation in the accuracy of the sound representations, additional variability can stem from language or musical (Tervaniemi, 2000) expertise and problems in auditory processing. For example, the inaccuracy in some feature(s) of the sound-event representation(s) can affect the recognition of the speech sounds from the acoustic stream. Backward masking might also, in some conditions such as dyslexia (Kujala et al., 2000) and chronic alcoholism (Ahveninen et al., 1999), be pathologically enhanced, thereby promoting the too rapid loss of sensory information contained by traces of preceding sound stimuli.

In conclusion, MMN provides an objective tool for studying how the auditory environment is represented in the human brain and how the conscious perception of sounds is achieved. The accuracy, development, and cortical locations of the sound representations can be precisely delineated with the MMN. Finally, MMN is becoming quite useful for testing cortical sound-discrimination accuracy and related attention/memory-switch functions in a variety of clinical populations.

ACKNOWLEDGMENTS

Figure 1 reprinted from Tiitinen, H., May, P., Reinikainen, K., & Näätänen, R. (1994). Attentive novelty detection in humans is governed by pre-attentive sensory memory. *Nature, 372,* 90-92. Copyright (2002), with permission from MacMillan Magazines Limited.

Figure 2 reprinted from Näätänen, R., & Alho, K. (1997). Mismatch negativity (MMN)— The measure for central sound representation accuracy. *Audiology & Neuro-Otology, 2,* 341-353. Copyright (2002), with permission from S. Karger AG, Basel.

Figure 4 reprinted from Näätänen, R., & Winkler, I. (1999). The concept of auditory stimulus representation in cognitive neuroscience. *Psychological Bulletin, 6,* 826-859. Copyright (2002), with permission from APA.

Figure 5 reprinted from Näätänen, R., Schröger, E., Karakas, S., Tervaniemi, M., & Paavilainen, P. (1993b). Development of a memory trace for a complex sound in the human brain. *NeuroReport, 4,* 503-506. Copyright (2002), with permission from Lippicott Williams & Wilkins.

Figure 6 reprinted from Näätänen, R., Lehtokoski, A., Lennes, M., Cheour, M., Huotilainen, M., Iivonen, A., Vainio, M., Alku, P., Ilmoniemi, R.J., Luuk, A., Allik, J., Sinkkonen, J., & Alho, K. (1997). Language-specific phoneme representations revealed by electric and magnetic brain responses. *Nature, 385,* 432-434. Copyright (2002), with permission from MacMillan Magazines Limited.

REFERENCES

Aaltonen, O., Eerola, O., Lang, H.A., Uusipaikka, E., & Tuomainen, J. (1994). Automatic discrimination of phonetically relevant and irrelevant vowel parameters as reflected by mismatch negativity. *Journal of the Acoustical Society of America, 3,* 1489-1493.

Aaltonen, O., Niemi, P., Nyrke, T., & Tuhkanen, J.M. (1987). Event-related brain potentials and the perception of a phonetic continuum. *Biological Psychology, 24,* 197-207.

Aaltonen, O., Tuomainen, J., Laine, M., & Niemi, P. (1993). Discrimination of speech and non-speech sounds by brain-damaged subjects: Electrophysiological evidence for distinct sensory processes. *Brain and Language, 44,* 139-152.

Ahveninen, J., Escera, C., Polo, M.D., Grau, C., & Jääskeläinen, I.P. (2000). Acute and chronic effects of alcohol on preattentive auditory processing as reflected by mismatch negativity. *Audiology & Neuro-Otology, 5,* 303-311.

Ahveninen, J., Jääskeläinen, I.P., Pekkonen, E., Hallberg, A., Hietanen, M., Mäkelä, R., Näätänen, R., & Sillanaukee, P. (1999). Suppression of mismatch negativity by backward masking predicts impaired working-memory performance in alcoholics. *Alcoholism: Clinical and Experimental Research, 23,* 1507-1515.

Alho, K. (1995). Cerebral generators of mismatch negativity (MMN) and its magnetic counterpart (MMNm) elicited by sound changes. *Ear and Hearing, 16,* 38-51.

Alho, K., Connolly, J.F., Cheour, M., Lehtokoski, A., Huotilainen, M., Virtanen, J., Aulanko, R., & Ilmoniemi, R. (1998). Hemispheric lateralization in preattentive processing of speech sounds. *Neuroscience Letters, 258,* 9-12.

Alho, K., Escera, C., Diaz, R., Yago, E., & Serra, J.M. (1997). Effects of involuntary auditory attention on visual task performance and brain activity. *NeuroReport, 8,* 3233-3237.

Alho, K., Huotilainen, M., Tiitinen, H., Ilmoniemi, R.J., Knuutila, J., & Näätänen, R. (1993). Memory-related processing of complex sound patterns in human auditory cortex: A MEG study. *NeuroReport, 4,* 391-394.

Alho, K., Sainio, K., Sajaniemi, N., Reinikainen, K., & Näätänen, R. (1990). Event-related brain potential of human newborns to pitch change of an acoustic stimulus. *Electroencephalography and Clinical Neurophysiology, 77,* 151-155.

Alho, K., Tervaniemi, M., Huotilainen, M., Lavikainen, J., Tiitinen, H., Ilmoniemi, R.J., Knuutila, J., & Näätänen, R. (1996). Processing of complex sounds in the human auditory cortex as revealed by magnetic brain responses. *Psychophysiology, 33,* 369-375.

Alho, K., Woods, D.L., Algazi, A., Knight, R.T., & Näätänen, R. (1994). Lesions of frontal cortex diminish the auditory mismatch negativity. *Electroencephalography and Clinical Neurophysiology, 91,* 353-362.

Aulanko, R., Hari, R., Lounasmaa, O.V., Näätänen, R., & Sams, M. (1993). Phonetic invariance in the human auditory cortex. *NeuroReport, 4,* 1356-1358.

Celsis, P., Boulanouar, K., Doyon, B., Ranjeva, J.P., Berry, I., Nespoulous, J.L., & Chollet, F. (1999). Differential fMRI responses in the left posterior superior temporal gyrus and left supramarginal gyrus to habituation and change detection in syllables and tones. *NeuroImage, 9,* 135-144.

Cheour, M., Ceponiene, R., Lehtokoski, A., Luuk, A., Allik, J., Alho, K., & Näätänen, R. (1998a). Development of language-specific phoneme representations in the infant brain. *Nature Neuroscience, 1,* 351-353.

Cheour, M., Haapanen, M-L., Ceponiene, R., Hukki, J., Ranta, R., & Näätänen, R. (1998b). Mismatch negativity (MMN) as an index of auditory sensory memory deficit in cleft-palate and CATCH syndrome children. *NeuroReport, 9,* 2709-2712.

Cheour-Luhtanen, M., Alho, K., Sainio, K., Rinne, T., Reinikainen, K., Pohjavuori, M., Aaltonen, O., Eerola, O., & Näätänen, R. (1996). The ontogenetically earliest discriminative response of the human brain. *Psychophysiology, Special Report, 33,* 478-481.

Cowan, N. (1988). Evolving conceptions of memory storage, selective attention, and their mutual constrains within the human information-processing system. *Psychological Bulletin, 104,* 163-191.

Cowan, N., Winkler, I., Teder, W., & Näätänen, R. (1993). Memory prerequisites of the mismatch negativity in the auditory event-related potential (ERP). *Journal of Experimental Psychology: Learning, Memory, and Cognition, 19,* 909-921.

Dehaene-Lambertz, G. (1997). Electrophysiological correlates of categorical phoneme perception in adults. *NeuroReport, 8,* 919-924.

Dehaene-Lambertz, G., & Baillet, S. (1998). A phonological representation in the infant brain. *NeuroReport, 9,* 1885-1888.

Dehaene-Lambertz, G., Dupoux, E., & Gout, A. (2000). Electrophysiological correlates of phonological processing: A cross-linguistic study. *Journal of Cognitive Neuroscience 12,* 635-647.

Deouell, L.Y., Hämäläinen, H., & Bentin, S. (2000). Unilateral neglect after right-hemisphere damage: Contributions from event-related potentials. *Audiology & Neuro-Otology, 5,* 225-234.

Escera, C., Alho, K., Schröger, E., & Winkler, I. (2000). Involuntary attention and distractibility as evaluated with event-related brain potentials. *Audiology & Neuro-Otology, 5,* 151-166.

Escera, C., Alho, K., Winkler, I., & Näätänen, R. (1998). Neural mechanisms of involuntary attention to acoustic novelty and change. *Journal of Cognitive Neuroscience, 10,* 590-604.

Fischer, C., Morlet, D., & Giard, M.H. (2000). Mismatch negativity and N100 in comatose patients. *Audiology & Neuro-Otology, 5,* 192-197.

Ford, J.M., & Hillyard, S.A. (1981). Event-related potentials, ERPs, to interruptions of steady rhythm. *Psychophysiology, 18,* 322-330.

Giard, M.-H., Lavikainen, J., Reinikainen, K., Perrin, F., Bertrand, O., Pernier, J., & Näätänen, R. (1995). Separate representation of stimulus frequency, intensity, and duration in auditory sensory memory: An event-related potential and dipole-model analysis. *Journal of Cognitive Neuroscience, 7,* 133-143.

Giard, M.-H., Perrin, F., Pernier, J., & Bouchet, P. (1990). Brain generators implicated in processing of auditory stimulus deviance: A topographic event-related potential study. *Psychophysiology, 27,* 627-640.

Gomes, H., Bernstein, R., Ritter, W., Vaughan, H.G., Jr., & Miller, J. (1997). Storage of feature conjunctions in transient memory. *Psychophysiology, 34,* 712-716.

Gootjes, L., Raij, T., Salmelin, R., & Hari, R. (1999). Left-hemisphere dominance for processing of vowels: A whole-scalp neuromagnetic study. *NeuroReport, 10,* 2987-2991.

Hari, R., Joutsiniemi, S.-L., Hämäläinen, M., & Vilkman, V. (1989). Neuromagnetic responses of human auditory cortex to interruptions in a steady rhythm. *Neuroscience Letters, 99,* 164-168.

Hawkins, H.L., & Presson, J.C. (1986). Auditory information processing. In K.R. Boff, L. Kaufman & J.P. Thomas (Eds.), *Handbook of perception and human performance* (pp. 61-64). New York: Wiley.

Ilvonen, T.-M., Kujala, T., Tervaniemi, M., Salonen, O., Näätänen, R., & Pekkonen, E. (2001). The processing of sound duration after left hemisphere stroke: Event-related potential and behavioral evidence. *Psychophysiology, 38,* 622-628.

Jääskeläinen, I.P., Hirvonen, J., Kujala, T., Alho, K., Erikson, C.J., Lehtokoski, A., Pekkonen, E., Sinclair, J.D., Yabe, H., & Näätänen, R. (1998). Effects of naltrexone and ethanol on auditory event-related brain potentials. *Alcohol, 15,* 105-111.

Jääskeläinen, I.P., Pekkonen, E., Hirvonen, J., Sillanaukee, P., & Näätänen, R. (1996). Mismatch negativity subcomponents and ethyl alcohol. *Biological Psychology, 43,* 13-25.

Kaipio, M.-L., Cheour, M., Ceponiene, R., Öhman, J., Alku, P., & Näätänen, R. (2000). Increased distractibility in closed head injury as revealed by event-related potentials. *NeuroReport, 11,* 1463-1468.

Kane, N.M., Curry, S.H., Butler, S.R., & Gummins, B.H. (1993). Electrophysiological indicator of awakening from coma. *Lancet, 341,* 688.

Kane, N.M., Curry, S.H., Rowlands, C.A., Manara, A.R., Lewis, T., Moss, T., Cummins, S.H., & Butler, S.R. (1996). Event-related potentials—neurophysiological tools for predicting emergence and early outcome from traumatic coma. *Intensive Care, 22,* 39-46.

Kaukoranta, E., Sams, M., Hari, R., Hämäläinen, M., & Näätänen, R. (1989). Reactions of human auditory cortex to changes in tone duration. *Hearing Research, 41,* 15-22.

Korpilahti, P., & Lang, H.A. (1994). Auditory ERP components and mismatch negativity in dysphasic children. *Electroencephalography and Clinical Neurophysiology, 91,* 256-264.

Korzyukov, O., Alho, K., Kujala, A., Gumenyuk, V., Ilmoniemi, R.J., Virtanen, J., Kropotov, J., & Näätänen, R. (1999). Electromagnetic responses of the human auditory cortex generated by sensory-memory based processing of tone-frequency changes. *Neuroscience Letters, 276,* 169-172.

Kraus, N., & Cheour, M. (2000). Speech-sound representation in the brain: Studies using mismatch negativity. *Audiology & Neuro-Otology, 5,* 140-150.

Kraus, N., McGee, T., Carrell, T., & Sharma, A. (1995). Neurophysiologic basis of speech discrimination. Mismatch negativity as an index of central auditory function. *Ear and Hearing,* 16, 19-37.

Kraus, N., McGee, T., Micco, A., Sharma, A., Carrell, T.D., & Nicol, T.G. (1993). Mismatch negativity in school-age children to speech stimuli that are just perceptibly different. *Electroencephalography and Clinical Neurophysiology, 88,* 123-130.

Kropotov, J.D, Alho, K., Näätänen, R., Ponomarev, V.A., Kropotova, O.G., Anichkov, A.D., & Nechaev, V.B. (2000). Human auditory-cortex mechanism of preattentive sound discrimination. *Neuroscience Letters, 280,* 87-90.

Kropotov, J.D., Näätänen, R., Sevostianov, A.V., Alho, K., Reinikainen, K., & Kropotova, O.V. (1995). Mismatch negativity to auditory stimulus change recorded directly from the human temporal cortex. *Psychophysiology, 32,* 418-422.

Kujala, T., Myllyviita, K., Tervaniemi, M., Alho, K., Kallio, J., & Näätänen, R. (2000). Basic auditory dysfunction in dyslexia as pinpointed by brain-activity measurements. *Psychophysiology, Special Report, 37,* 262-266.

Lang, H., Nyrke, T., Ek, M., Aaltonen, O., Raimo, I., & Näätänen, R. (1990). Pitch discrimination performance and auditory event-related potentials. In C.H.M. Brunia, A.W.K. Gaillard, A. Kok, G. Mulder, & M.N. Verbaten (Eds.), *Psychophysiological brain research, Vol. 1* (pp. 294-298). Tilburg: Tilburg University Press.

Leppänen, P.H.T., & Lyytinen, H. (1997). Auditory event-related potentials in the study of developmental language-related disorders. *Audiology & Neuro-Otology, 2,* 308-340.

Levänen, S., Hari, R., McEvoy, L., & Sams, M. (1993). Responses of the human auditory cortex to changes in one versus two stimulus features. *Experimental Brain Research, 97,* 177-183.

Lounasmaa, O.V., Hari, R., Joutsiniemi, S-L., & Hämäläinen, M. (1989). Multi-SQUID recordings of human cerebral magnetic fields may give information about memory processes. *Europhysics Letters, 9,* 603-608.

Lyytinen, H., Blomberg, A.-P., & Näätänen, R. (1992). Event-related potentials and autonomic responses to a change in unattended auditory stimuli. *Psychophysiology, 29,* 523-534.

Morlet, D., Bouchet, P., & Fischer, C. (2000). Mismatch negativity and N100 monitoring: potential clinical value and methodological advances. *Audiology & Neuro-Otology, 5,* 198-206.

Näätänen, R. (1985). Selective attention and stimulus processing: Reflections in event-related potentials, magnetoencephalogram and regional cerebral blood flow. In M.I. Posner & O.S.M. Marin (Eds.), *Attention and performance XI* (pp. 355-373). Hillsdale, NJ: Erlbaum.

Näätänen, R. (1990). The role of attention in auditory information processing as revealed by event-related potentials and other brain measures of cognitive function. *The Behavioral and Brain Sciences, 13,* 201-288.

Näätänen, R. (1991). Mismatch negativity (MMN) outside strong attentional focus: A commentary on Woldorff et al. *Psychophysiology, 28,* 478-484.

Näätänen, R. (1992). *Attention and brain function.* Hillsdale, NJ: Lawrence Erlbaum Associates.

Näätänen, R. (2001). The perception of speech sounds by the human brain as reflected by the mismatch negativity (MMN) and its magnetic equivalent (MMNm). *Psychophysiology, 38,* 1-21

Näätänen, R., & Alho, K. (1995). Mismatch negativity—A unique measure of sensory processing in audition. *Ear and Hearing, 16,* 6-18.

Näätänen, R., & Alho, K. (1997). Mismatch negativity (MMN)—The measure for central sound representation accuracy. *Audiology & Neuro-Otology, 2,* 341-353.

Näätänen, R., Gaillard, A., & Mäntysalo, S. (1978). Early selective-attention effect on evoked potential reinterpreted. *Acta Psychologica, 42,* 313-329.

Näätänen, R., Jiang, D., Lavikainen, J., Reinikainen, K., & Paavilainen, P. (1993a). Event-related potentials reveal a memory trace for temporal features. *NeuroReport, 5,* 310-312.

Näätänen, R., Lehtokoski, A., Lennes, M., Cheour, M., Huotilainen, M., Iivonen, A., Vainio, M., Alku, P., Ilmoniemi, R.J., Luuk, A., Allik, J., Sinkkonen, J., & Alho, K. (1997). Language-specific phoneme representations revealed by electric and magnetic brain responses. *Nature, 385,* 432-434.

Näätänen, R., Paavilainen P., Alho, K., Reinikainen K., & Sams, M. (1987a). The mismatch negativity to intensity changes in an auditory stimulus sequence. In R. Johnson, Jr., J.W. Rohrbaugh, & R. Parasuraman (Eds.), *Current trends in event-related potential research* (pp. 125-131). Amsterdam: Elsevier.

Näätänen, R., Paavilainen P., Alho, K., Reinikainen K., & Sams, M. (1987b). Inter-stimulus interval and the mismatch negativity. In C. Barber & T. Blum (Eds.), *Evoked potentials III* (pp. 392-397). London: Butterworths.

Näätänen, R., Paavilainen P., Alho, K., Reinikainen K., & Sams, M. (1989a). Do event-related potentials reveal the mechanism of the auditory sensory memory in the human brain? *Neuroscience Letters, 98,* 217-221.

Näätänen, R., Paavilainen P., & Reinikainen K. (1989b). Do event-related potentials to infrequent decrements in duration of auditory stimuli demonstrate a memory trace in man? *Neuroscience Letters, 107,* 347-352.

Näätänen, R., Schröger, E., Karakas, S., Tervaniemi, M., & Paavilainen, P. (1993b). Development of a memory trace for a complex sound in the human brain. *NeuroReport, 4,* 503-506.

Näätänen, R., & Winkler, I. (1999). The concept of auditory stimulus representation in cognitive neuroscience. *Psychological Bulletin, 6,* 826-859.

Nordby, H., Hammerborg, D., Roth, W.T., & Hughdal, K. (1988). Event-related potentials to time-deviant and pitch-deviant tones. *Psychophysiology, 25,* 249-261.

Ogura, C., Nageihsi, Y., Omura, F., Fukao, K., Ohta, H., Kishimoto, A., & Matsubayashi, M. (1993). N200 component of the event-related potentials in depression. *Biological Psychiatry, 33,* 720-726.

Paavilainen, P., Alho, K., Reinikainen, K., Sams, M., & Näätänen, R. (1991). Right-hemisphere dominance of different mismatch negativities. *Electroencephalography and Clinical Neurophysiology, 78,* 466-479.

Paavilainen, P., Karlsson, M-L., Reinikainen, K., & Näätänen, R. (1989). Mismatch negativity to change in the spatial location of an auditory stimulus. *Electroencephalography and Clinical Neurophysiology, 73,* 129-141.

Pekkonen, E. (2000). Mismatch negativity in aging, and in Alzheimer's and Parkinson's diseases. *Audiology & Neuro-Otology, 5,* 216-224.

Phillips, C., Pellathy, T., Marantz, A., Yellin, E., Wexler, K., Poeppel, D., McGinnis, M., & Roberts, T. (2000*).* Auditory cortex accesses phonological categories: An MEG mismatch study. *Journal of Cognitive Neuroscience 12,* 1038-1055.

Polo, M.D., Escera, C., Gual, A., & Grau, C. (1999). Mismatch negativity and auditory sensory memory in chronic alcoholics. *Alcoholism: Clinical & Experimental Research, 23,* 1744-50.

Ponton, C.W., Eggermont, J.J., Don, M., Waring, M.D., Kwong, B., Cunningham, J., & Trautwein, P. (2000). Maturation of the mismatch negativity: Effects of profound deafness and cochlear implant use. *Audiology & Neuro-Otology, 5,* 167-185.

Pulvermüller, F., Kujala, T., Sthyrov, Y., Simola, J., Tiitinen, H., Alku, P., Alho, K., Ilmoniemi, R.J., & Näätänen, R. (2001). Memory traces for words as revealed by the mismatch negativity (MMN). *NeuroImage, 14,* 607-616.

Rinne, T., Alho, K., Alku, P., Holi, M., Sinkkonen, J., Virtanen, J., Bertrand, O., & Näätänen, R. (1999). Analysis of speech sounds is left-hemisphere predominant at 100-150 ms after sound onset. *NeuroReport, 10,* 1113-1117.

Rinne, T., Alho, K., Ilmoniemi, R.J., Virtanen, J., & Näätänen, R. (2000). Separate time behaviors of the temporal and frontal mismatch negativity sources. *NeuroImage, 12,* 14-19.

Ritter, W., Deacon, D., Gomes, H., Javitt, D.C., & Vaughan, H.G, Jr. (1995). The mismatch negativity of event-related potentials as a probe of transient auditory memory: A review. *Ear and Hearing, 16,* 52-67.

Sams, M., Alho, K., & Näätänen, R. (1984). Short-term habituation and dishabituation of the mismatch negativity on the ERP. *Psychophysiology, 21,* 434-441.

Sams, M., Hari, R., Rif, J., & Knuutila, J. (1993). The human auditory sensory memory trace persists about 10s: Neuromagnetic evidence. *Journal of Cognitive Neuroscience, 5,* 363-370.

Sams, M., Paavilainen, P., Alho, K., & Näätänen, R. (1985). Auditory frequency discrimination and event-related potentials. *Electroencephalography and Clinical Neurophysiology, 62,* 437-448.

Scharf, B., & Houstma, A.J. (1986). Audition II: Loudness, pitch, localization, aural distortion, pathology. In K.R. Boff, L. Kaufman, & J.P. Thomas (Eds.), *Handbook of perception and human performance* (pp. 15.1-15.60). New York: Wiley.

Schröger, E. (1994). An event-related potential study of sensory representations of unfamiliar tonal patterns. *Psychophysiology, 31,* 175-181.

Schröger, E. (1996). A neural mechanism for involuntary attention shifts to changes in auditory stimulation. *Journal of Cognitive Neuroscience, 8,* 527-539.

Schröger, E. (1997). On the detection of auditory deviants: Pre-attentive activation model. *Psychophysiology, 34,* 245-257.

Schröger, E., & Wolff, C. (1996). Mismatch response of the human brain to changes in sound location. *NeuroReport, 7,* 3005-3008.

Sharma, A., & Kraus, N. (1995). Effects of contextual variations in pitch and phonetic processing: Neurophysiologic correlates. *Association for Research in Otolaryngology, 729,* 183.

Shelley, A.M., Ward, P.B., Catts, S.V., Michie, P.T., Andrews, S., & McConaghy, N. (1991). Mismatch negativity: an index of a preattentive processing deficit in schizophrenia. *Biological Psychiatry, 30,* 1059-1062.

Shtyrov, Y., Kujala, T., Ahveninen, J., Tervaniemi, M., Alku, P., Ilmoniemi, R.J., & Näätänen, R. (1998). Background acoustic noise and the hemispheric lateralization of speech processing in the human brain: magnetic mismatch negativity study. *Neuroscience Letters, 251,* 141-144.

Shtyrov, Y., Kujala, T., Lyytinen, H., Kujala, J., Ilmoniemi, R.J., & Näätänen, R. (2000). Lateralization of speech processing in the brain as indicated by mismatch negativity and dichotic listening. *Brain and Cognition, 43,* 392-398.

Sussman, E., Gomes, H., Nousak, J.M., Ritter, W., & Vaughan, H.G., Jr. (1998). Feature conjunctions and auditory sensory memory. *Brain Research, 18,* 95-102.

Sussman, E., Winkler, I., Ritter, W., Alho, K., & Näätänen, R. (1999). Temporal integration of auditory stimulus deviance as reflected by the mismatch negativity. *Neuroscience Letters, 264,* 161-164.

Takegata, R., Paavilainen, P., Näätänen, R., & Winkler, I. (1999). Independent processing of changes in auditory single features and feature conjunctions in humans as indexed by the mismatch negativity (MMN). *Neuroscience Letters, 266,* 109-112.

Tervaniemi, M. (2000). Automatic processing of musical information as evidenced by EEG and MEG recordings. In T. Nakada (Ed.), *Integrated human brain science: Theory, method and application (Music)* (pp. 325-335). Amsterdam: Elsevier Science B.V.

Tervaniemi, M., Ilvonen, T., Karma, K., Alho, K., & Näätänen, R. (1997). The musical brain: Brain waves reveal the neurophysiological basis of musicality. *Neuroscience Letters, 226,* 1-4.

Tervaniemi, M., Kujala, A., Alho, K., Virtanen, J., Ilmoniemi, R.J., & Näätänen, R. (1999). Functional specialization of the human auditory cortex in processing phonetic and musical sounds: A magnetoencephalographic (MEG) study. *NeuroImage, 9,* 330-336.

Tervaniemi, M., Maury, S., & Näätänen, R. (1994). Neural representations of abstract stimulus features in the human brain as reflected by the mismatch negativity. *NeuroReport, 5,* 844-846.

Tervaniemi, M., Medvedev, S.V., Alho, K., Pakhomov, S.V., Roudas, M.S., van Zuijen, T.L., & Näätänen, R. (2000). Lateralized automatic auditory processing of phonetic versus musical information: A PET study. *Human Brain Mapping, 10,* 74-79.

Tiitinen, H., May, P., Reinikainen, K., & Näätänen, R. (1994). Attentive novelty detection in humans is governed by pre-attentive sensory memory. *Nature, 372,* 90-92.

Trejo, L.J., Ryan Jones, D.L., & Kramer, A.F. (1995). Attentional modulation of the mismatch negativity elicited by frequency differences between binaurally presented tone bursts. *Psychophysiology, 32,* 319-328.

Winkler, I., Cowan, N., Csépe, V., Czigler, I., & Näätänen, R. (1996a). Interactions between transient and long-term auditory memory as reflected by the mismatch negativity. *Journal of Cognitive Neuroscience, 8,* 403-415.

Winkler, I., Czigler, I., Jaramillo, M., Paavilainen, P., & Näätänen, R. (1998). Temporal constraints of auditory event synthesis: Evidence from ERPs. *NeuroReport, 9,* 495-499.

Winkler, I., Karmos, G., & Näätänen, R. (1996b). Adaptive modeling of the unattended acoustic environment reflected in the mismatch negativity event-related potential. *Brain Research, 742,* 239-252.

Winkler, I., Kujala, T., Tiitinen, H., Sivonen, P., Alku, P., Lehtokoski, A., Czigler, I., Csépe, V., Ilmoniemi, R.J., & Näätänen, R. (1999a). Brain responses reveal the learning of foreign language phonemes. *Psychophysiology, 36,* 638-642.

Winkler, I., Lehtokoski, A., Alku, P., Vainio, M., Czigler, I., Csépe, V., Aaltonen, O., Raimo, I., Alho, K., Lang, H., Iivonen, A., & Näätänen, R. (1999b). Pre-attentive detection of vowel contrasts utilizes both phonetic and auditory memory representations. *Cognitive Brain Research, 7,* 357-369.

Winkler, I., & Näätänen, R. (1992). Event-related potentials in auditory backward recognition masking: A new way to study the neurophysiological basis of sensory memory in humans. *Neuroscience Letters, 140,* 239-242.

Winkler, I., & Schröger, E. (1995). Neural representation for the temporal structure of sound patterns. *NeuroReport, 6,* 690-694.

Winkler, I., Tervaniemi, M., & Näätänen, R. (1997). Two separate codes for missing fundamental pitch in the human auditory cortex. *Journal of the Acoustical Society of America, 102,* 1072-1082.

Woldorff, M.G., Hackley, S.A., & Hillyard, S.A. (1991). The effects of channel-selective attention on the mismatch negativity wave elicited by deviant tones. *Psychophysiology, 28,* 30-42.

Yabe, H., Tervaniemi, M., Reinikainen, K., & Näätänen, R. (1997). Temporal window of integration revealed by MMN to sound omission. *NeuroReport, 8,* 1971-1974.

Yabe, H., Tervaniemi, M., Sinkkonen, J., Huotilainen, M., Ilmoniemi, R.J., & Näätänen, R. (1998). Temporal window of integration of auditory information in the human brain. *Psychophysiology, 35,* 615-619.

Chapter 2

EVENT-RELATED BRAIN POTENTIAL INDICES OF INVOLUNTARY ATTENTION TO AUDITORY STIMULUS CHANGES

KIMMO ALHO, CARLES ESCERA, AND ERICH SCHRÖGER
Cognitive Brain Research Unit, Department of Psychology, University of Helsinki, Finland, Neurodynamics Laboratory, Department of Psychiatry and Clinical Psychobiology, University of Barcelona, Catalonia-Spain, and Institute for General Psychology, University of Leipzig, Germany

As first reported by Näätänen, Gaillard, and Mäntysalo (1978, 1980), infrequent ("deviant") sounds occurring in a sequence of repetitive ("standard") sounds elicit the mismatch negativity (MMN) event-related brain potential (ERP), even when the listener is instructed to attend to other stimuli. MMN is seen as a negative-polarity displacement of the ERP to deviant sounds in relation to the ERP from standard sounds around 100-200 milliseconds from deviant-event onset (see Figure 1). As in the early reports of Näätänen and his colleagues, most subsequent MMN studies have applied tones as stimuli and deviancies in some simple feature (e.g., pitch, intensity, duration, or location) to elicit MMN (for a review, see Näätänen, 1992). However, MMN is also elicited by infrequent changes and irregularities in complex sounds, such as phonemes, syllables, chords, and tone patterns (for recent reviews, see Näätänen & Alho, 1997; Näätänen & Winkler, 1999; Schröger, 1997).

The brain process that generates MMN is evidently triggered by a mismatch between a deviant auditory stimulus or event and a memory representation of the regularities in the preceding auditory stimulation (Näätänen, 1992; Winkler et al., 1996; Winkler & Czigler, 1998). This interpretation is supported by results showing that infrequent sounds presented alone, without intervening repetitive sounds or a change in the beginning of a sound sequence, do not elicit the MMN (Cowan et al., 1993; Korzyukov et al., 1999; Kropotov et al., 2000; Näätänen et al., 1989).

MMN has proven to be a successful tool for studying preattentive auditory perception and memory functions (Näätänen, 1995; Näätänen & Alho, 1997; Näätänen & Winkler, 1999; Ritter et al., 1995; Schröger, 1997). For example, the speed of the preattentive processing of auditory stimulus changes that generates the MMN appears to explain the speed of active discrimination of these changes when the auditory stimuli are attended, as shown by the strong correlation between the decrease of MMN peak latency and shortening of reaction times (RTs) with increasing stimulus deviance (Tiitinen et al., 1994). Such active discrimination of stimulus changes is reflected by the N2b, another negative ERP component, which overlaps MMN and is followed by the positive P3b (P300) component (Alho et al., 1992; Donchin & Coles, 1988; Näätänen, 1990, 1992; Näätänen et al., 1993; Sams et al., 1985; Sutton et al., 1965). However, the functional role of the brain process generating the MMN is likely associated with orienting response initiation (Öhman, 1979; Sokolov, 1975) to changes in the auditory environment (Näätänen, 1992; Näätänen & Michie, 1979; Schröger, 1996).

Figure 2 illustrates the fronto-centrally dominant scalp distribution of MMN, which is mainly explained by the sum of bilaterally generated auditory-cortex activity in these brain areas (Giard et al., 1995; Rinne et al. 1999a; Scherg et al., 1989). Source modeling of MMNm, the magnetoencephalographic (MEG) counterpart of MMN, supports this interpretation (Alho et al., 1998; Hari et al., 1984; Levänen et al., 1996). Intracranial MMN recordings have also indicated MMN generation occurs in the auditory cortex (Csépe et al., 1987; Halgren et al., 1995a; Javitt et al., 1996; Kraus et al., 1994; Kropotov et al., 1995, 2000; Liasis et al., 1999). Additionally, studies applying functional magnetic resonance imaging (fMRI, Celsis et al., 1999; Opitz et al., 1999), positron emission tomography (PET, Tervaniemi et al., 2000), and event-related optical signals are also consistent with this source of MMN generation. (EROS, Rinne et al., 1999b). Moreover, patients with temporal-cortex lesions demonstrated attenuated scalp-recorded MMNs (Aaltonen et al., 1993; Alain et al., 1998).

However, scalp current density (SCD) analysis of MMN voltage distribution over the head suggests that MMN gets an additional contribution from the prefrontal brain areas (Deouell et al., 1998; Giard et al., 1990; Rinne et al., 2000; Serra et al., 1998; Yago et al., 2001). This finding is supported by MEG (Levänen et al., 1996) and fMRI recordings (Celsis et al., 1999; Opitz et al., 2002). As prefrontal cortex has an important role in determining the direction of attention (Fuster, 1986; Stuss & Benson, 1989), the prefrontal MMN activity is thought to index the involuntary orienting of attention to a change in the acoustic environment detected by the auditory-cortex MMN mechanism (Giard et al., 1990; Näätänen 1992; Näätänen & Michie, 1979).

Figure 1 also illustrates the positive P3a response that often follows the MMN (Alho et al., 1998; Escera et al., 1998; Näätänen et al., 1982; Sams et al., 1985), which is associated with the actual switching of attention (Escera et al., 1998; Ford et al., 1976; Knight & Scabini, 1998; Squires et al., 1975; Woods, 1990). This association is suggested by results showing that large P3a responses are elicited by attention-catching, widely deviant, complex "novel" sounds (e.g., a dog barking or a telephone ringing) even when they occur in a to-be-ignored auditory stimulus sequence (Alho et al., 1998; Escera et al., 1998; Woods, 1992; Woods et al., 1993). When a small change occurs in an unattended auditory stimulus sequence, P3a in the ERP of a deviant stimulus may be quite small in amplitude or may fail to be elicited, perhaps because small stimulus changes do not always catch attention (Alho et al., 1992, 1998; Escera et al., 1998; Sams et al., 1985; Tiitinen et al., 1994). This lack of response is indicated by results showing that even when such changes elicit MMN, they are not always followed by heart-rate deceleration or skin-conductance increase controlled by the autonomic nervous system and commonly associated with involuntary orienting of attention (Lyytinen et al., 1992). According to SCD mapping of P3a (Yago et al., submitted), source modeling of P3a (Mecklinger & Ullsperger, 1995) and its MEG counterpart of P3a (Alho et al., 1998), intracranial P3a recordings (Alain et al., 1989; Baudena et al., 1995; Halgren et al., 1995a,b; Kropotov et al., 1995), and effects of local brain lesions on P3a (Knight, 1984, 1996; Knight et al., 1989), the distributed cerebral network of involuntary attention switching activated by novel sounds includes at least the superior temporal, dorsolateral prefrontal, and parietal cortical areas, the parahippocampal and anterior cingulate gyri, and the hippocampus.

However, infrequent auditory stimuli occurring outside the current focus of attention may cause involuntary attention switching without eliciting MMN, as is the case for infrequent sounds delivered without intervening standard sounds and for sounds beginning an auditory stimulus sequence after a relatively long silent period (Näätänen, 1992). These sounds do not elicit MMN but they evoke an enhanced N1 component (Cowan et al., 1993; Kropotov et al., 2000; Korzyukov et al., 1999; Näätänen et al., 1989), which peaks at about 100 millisecond from stimulus onset and is sensitive to stimulation rate (Näätänen & Picton, 1987). A sound repeated at a high rate elicits only a small N1, whereas sounds delivered at a low rate elicit a large N1, due to enhanced activity of modality-specific and non-specific brain areas (Giard et al., 1994; Hari et al., 1982; Näätänen & Picton, 1987; Näätänen & Winkler, 1999). Moreover, in addition to MMN, a widely deviant sound in a sequence of repeating sounds (e.g., a novel sound among tone pips) may elicit an enhanced N1, presumably because it activates a population of new feature-specific (e.g., frequency-specific) neurons in the auditory cortex (Alho et al., 1998; Escera et al., 1998). It has been suggested that although MMN is generated by a process initiating involuntary attention to auditory stimulus changes, the auditory N1 response is generated by a process involved in directing the focus of attention to onsets of new events in the auditory environment (Näätänen, 1992). This suggestion is supported by findings that novel sounds that elicit an enhanced N1 also elicit a large P3a, indicating engagement of attention by these sounds (Alho et al., 1998; Escera et al., 1998; Woods, 1990). Finally, ERPs also provide an index for redirecting attention back to the current task after involuntary switching of attention away from this task. This redirecting process might generate the recently discovered "reorienting negativity" (RON; Schröger & Wolff, 1998a) following P3a response to deviant stimuli, as shown in Figure 1.

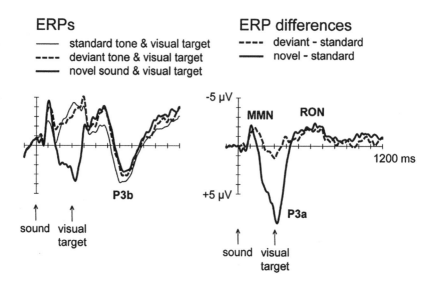

Figure 1. Event-related brain potentials (ERPs) recorded at the central midline scalp site (Cz) to task-irrelevant standard tones of 600 Hz, to 700-Hz deviant tones, and complex novel sounds infrequently replacing the standard tone (sound onset at 0 millisecond). Each sound was followed by a task-relevant visual stimulus (onset at 300 milliseconds) that was to be discriminated and responded to by the subject and therefore elicited the positive P3b component. Right: Difference waves obtained by subtracting the ERP following the standard tones from the ERP following deviant tones and from the ERP following novel sounds. The difference wave for deviant tones shows the mismatch negativity (MMN), followed by a small positive P3a response and the "reorienting negativity" (RON). The difference wave for novel sounds shows a negative wave, consisting of MMN and an enhanced N1 response, followed by a large P3a and by RON (after Escera et al., 1998).

MMN
150 ms

Figure 2. Voltage distribution map for the fronto-centrally maximal mismatch negativity (MMN) elicited by infrequent, slightly higher deviant tones (700 Hz) occurring among repetitive standard tones (600 Hz) presented to a subject concentrating on a visual task. MMN amplitudes were measured around the MMN peak latency (150 milliseconds after stimulus onset) from difference waves obtained by subtracting event-related potentials (ERP) to standard tones at different scalp sites from those to deviant tones (cf. Figure 1). The head is viewed from above, the nose pointing upwards. Lighter shades of gray indicate more negative voltages and the small circles indicate the locations of scalp electrodes used to record MMN (after Yago et al., 2001).

1. EVENT-RELATED BRAIN POTENTIALS TO SOUND CHANGES DISTRACTING AUDITORY TASK PERFORMANCE

Evidence for the involvement of the MMN and P3a generator mechanisms in involuntary attention switching has been provided by simultaneous registration of behavioral and ERP effects of auditory distractors. Measurements of ERPs to auditory stimulus changes and distraction of auditory task performance by these changes have been carried out with two different types of tasks. In two-channel selective-attention tasks, subjects are dichotically presented with sounds and instructed to selectively attend to the input of one ear and to respond to particular target sounds occurring in this input. Distraction is indicated by poorer performance when target stimuli are preceded by infrequent, deviant sounds in the unattended input relative to the performance when the targets are preceded by frequent, standard sounds in the attended or unattended input. In one-channel tasks, distracting and task-relevant aspects of stimulation are embedded in the same acoustic event. For example, subjects are presented with equiprobably occurring short and long sounds that can be of a frequent standard frequency or of an infrequent deviant frequency. The subjects' task is to discriminate short from long tones and to disregard task-irrelevant frequency changes. Distraction is indicated by poorer duration-discrimination performance when a task-irrelevant frequency change occurs in the sound.

For example, Woods and colleagues (1993) used a two-channel selective-attention task in which high-frequency tones were presented to one ear and low-frequency tones to the other in a random order. Infrequent long-duration tones and occasional novel sounds were randomly interspersed in the stimulus sequence. Subjects were instructed to attend to tones delivered to one ear and to make a button-press response to long-duration tones occurring among the attended tones. RTs to these target stimuli were prolonged by more than 300 milliseconds when a target was preceded by a task-irrelevant novel sound even when the preceding novel sound occurred among the to-be-ignored tones. Novel sounds elicited MMNs followed by N2b and P3a responses both when they occurred among the to-be-attended tones and when they occurred among the to-be-ignored tones. No effects of duration deviants in the unattended channel on subsequent targets in the attended channel were reported. Longer-duration deviant tones in the attended and unattended input elicited MMNs, but the N2b and P3 responses were confined to duration-deviants (targets) in the attended input. This pattern of results suggests that sound changes in the unattended channel may cause involuntary switches of attention, but these changes have to be rather salient in order to become effective distractors.

In another two-channel study (Schröger, 1996), pairs of tone stimuli (S1 and S2) were presented, with subjects instructed to ignore S1 (delivered to the left ear) and to make a go/no-go response to the subsequent S2 (delivered to the right ear). On most trials, the task-irrelevant S1 was of standard frequency, but occasionally its frequency deviated slightly or widely from the standard frequency. Deviant tones elicited MMN followed by a small P3a. Distraction of the processing of task-relevant S2 by a deviant S1 was associated with prolonged RTs and attenuated N1 ERPs to S2 tones when preceded by a deviant S1 tone. These results support a model according to which the auditory system possesses a change detection system that monitors the acoustic input and may produce an attentional "interrupt" signal when a deviant occurs. However, this involuntary capture of attention caused by a deviant sound is rather short-lived, since distraction was confined to short S1-S2 intervals (200 milliseconds) and did not occur with long ones (560 milliseconds).

In one-channel distraction studies, even slight irrelevant physical changes in task-relevant sounds were found to cause large distraction effects. For example, in studies by Schröger and Wolff (1998a, b), subjects were to discriminate the duration of equiprobable tones that were of short (200 milliseconds) or long (400 milliseconds) duration. Tones were of a frequent standard frequency (700 Hz) or infrequent deviant frequency (750 Hz), and this frequency variation had no task relevance. RTs in the duration-discrimination task were prolonged for deviant-frequency tones compared with standard-frequency tones. ERPs to deviant-frequency tones showed MMN and P3a to these tones. They also elicited similar MMNs in another condition in which attention was directed away from the sounds indicating preattentive registration of the deviant sound. However, these small frequency changes elicited no P3a when the sounds were not attended (Schröger et al., 2000; Schröger & Wolff, 1998b). It seems likely that when the tones were totally task-irrelevant, it was possible to ignore small frequency changes in them, as indicated by the lack of P3a to these changes. However, when the tones carried task-relevant sound information, task-irrelevant sound information could not be easily disregarded, as indicated by P3a to the small frequency deviances during the tone-duration discrimination task, suggesting that the frequency changes caught the listeners' attention.

In the one-channel experiments of Schröger and colleagues, the P3a response to frequency changes was usually followed by a fronto-centrally distributed negativity at the 400-600 milliseconds range, as shown in Figure 1. However, this negativity was confined to conditions in which subjects discriminated long sounds from short ones and did not occur when the sounds were ignored or when the deviation was made task-relevant. Therefore, it seems likely this negativity is associated with reorienting of attention towards the task-relevant aspects of stimulation following distraction (Schröger et al., 2000; Schröger & Wolff, 1998b). For this reason, this negativity was named the "reorienting negativity" (RON). SCD maps for RON reveal bilateral frontal sinks around the fronto-central FC1 and FC2 electrode sites, suggesting frontal generators and complex current fields of lower amplitudes over centro-parietal regions (Schröger et al., 2000). Interestingly, RON seems not to be modality-specific as it was observed even in an analogous visual paradigm (Berti & Schröger, 2001), as shown in Figure 3, and even in a cross-modal –auditory-visual distraction paradigm (Escera et al., 2001; see below).

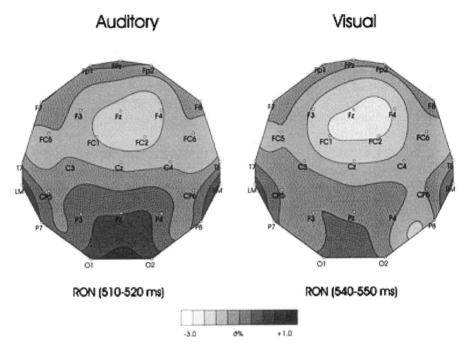

Figure 3. Voltage distribution map for the frontally maximal reorienting negativity (RON) following an occurrence of deviant-frequency tone. Subjects were to discriminate between equiprobable short (200 milliseconds) and long (400 milliseconds) tones that were either of a repetitive standard frequency (1000 Hz) or, infrequently, of a deviant frequency (950 or 1050 Hz). RON amplitudes were measured at 510-520 milliseconds after tone onset from difference waves obtained by subtracting event-related potentials (ERPs) following standard-frequency tones from those following deviant-frequency tones (Figure 1). For other details, see Figure 2. Right: The same as at the left for RON following a deviant visual stimulus. Subjects were to discriminate between equiprobable visual stimuli presented for a short (200 milliseconds) or long (600 milliseconds) time. Frequent standard stimuli were green squares containing a gray triangle and infrequent deviant stimuli were similar squares, but with a mislocated triangle. RON amplitudes were measured at 540-550 milliseconds after figure onset from deviant-standard ERP difference waves (after Berti & Schröger, 2001).

It is important to emphasize that the one-channel distraction task yields reliable distraction effects even with very small difference between deviant and standard sounds. So far, each subject studied with this paradigm showed a behavioral distraction effect (Jääskeläinen et al., 1999; Schröger & Berti, 2000; Schröger et al., 2000; Schröger & Wolff, 1998a, 1998b). The behavioral and electrophysiological effects observed in this distraction paradigm are highly replicable. The product-moment correlations for MMN, P3a, and RON and the RT prolongation measured in two separate sessions were between 0.77 and 0.90 (Schröger et al., 2000). In a very recent experiment (Schröger & Wolff, in preparation), the magnitude of frequency change was manipulated in five steps to determine the smallest frequency change still yielding behavioral distraction. The standard tone was of 1000 Hz and the deviant tones were 0.5%, 1.5%, 2.5%, 3.5% and 4.5% lower or higher in frequency. The 0.5% deviant did not cause a distraction effect. However, with a difference of only 1.5 %, the distraction effect produced a prolongation of RT that was about 20 milliseconds, with all 12 subjects showing this effect. The effect was of similar magnitude in the 2.5% and 3.5% conditions and somewhat larger in the 4.5% condition (32 milliseconds); each subject in each of these three conditions showed the RT effect (except in the 2.5% condition where one subject did not). These results suggest that frequency deviation may become perturbing as soon as it is above the discrimination threshold. The finding that MMN, P3a, and RON developed together with the behavioral distraction effect provides further evidence for the hypothesis that these components are indeed functionally related to distraction observed at the behavioral level.

2. EVENT-RELATED BRAIN POTENTIALS TO SOUND CHANGES DISTRACTING VISUAL TASK PERFORMANCE

The involvement of the MMN and the P3a generator processes in involuntary attention may also be observed cross-modally as shown by Escera and colleagues (1998). In their study, subjects were instructed to discriminate between two categories of visual stimuli (odd and even numbers) presented in a random order at a constant rate (1 stimulus in 1.2 seconds) on a computer screen, and to press the corresponding response button as fast and accurately as possible. A task-irrelevant sound occurred 300 milliseconds before each visual stimulus. The sound was either a frequently occurring (600 Hz, p=0.8) standard tone, a slightly higher deviant tone (700 Hz, p=0.1), or a novel sound (p=0.1) drawn from a pool of complex environmental sounds. Figure 4 illustrates the RTs to visual stimuli following deviant tones and novel sounds that were about 5 to 20 milliseconds longer, respectively, in comparison to RT to visual stimuli that followed standard tones. Unexpectedly, the hit rate was similar after standard tones and novel sounds but significantly reduced (by about 2%) after deviant tones, which resulted from an increased number of wrong responses to the visual stimuli following deviant tones (Figure 4). These effects were replicated in several studies using the same or slightly modified versions of this auditory-visual distraction paradigm (Alho et al., 1997; Alho et al., 1999; Escera et al., 1998, 2001, in press; Jääskeläinen et al., 1996; Yago et al., 2001, submitted).

Reaction Time

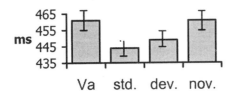

Hit Rate

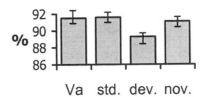

Error Rate

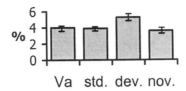

Figure 4. Performance speed and accuracy in the auditory-visual distraction paradigm showing that reaction times (RTs) are longer to to-be-discriminated visual stimuli following task-irrelevant frequency-deviant tones (DEV) and novel sounds (NOV) than to visual stimuli following standard tones (STD). Deviant tones are also followed by a hit-rate decrease in the visual performance due to an increased number of wrong responses. When the sounds were omitted and the visual stimuli were presented alone in a control condition (V), RTs to visual stimuli were slower than for the visual stimuli preceded by a standard tone. This outcome suggests that the sounds were not totally task-irrelevant but served as timing cues for visual stimuli and therefore speeded up the visual-task performance in the auditory-visual condition (after Escera et al., 1998).

Figure 4 also indicates that the RTs were shorter to the visual stimuli preceded by a standard tone than to similar visual stimuli in a control condition, in which the sounds were omitted from the stimulus sequence (Escera et al., 1998). This finding indicates that when the sounds were present, subjects covertly monitored the task-irrelevant auditory stimuli and used them as warning signals for the occurrence of the visual targets thereby speeding performance. Therefore, it may be argued that the observed effects of sound change on visual task performance did not truly reflect involuntary attention, as the subjects were covertly attending to the sounds. However, similar effects were observed in a related experiment by Alho et al. (1997) in which the subject's attention was more effectively directed away from the auditory stimuli and an occurrence of an auditory stimulus did not predict the presence of the subsequent task-relevant visual stimulus. This situation was achieved by presenting with each sound (either a standard or deviant tone) a simultaneous visual warning cue informing the subject of whether a successive task-relevant stimulus would be delivered or not. As in the Escera et al. (1998) study, prolonged RTs and increased error rates to visual stimuli following task-irrelevant deviant tones were observed, thereby confirming the involuntary nature of the attention capture by the deviant tones. Moreover, it was found that deviant tones preceding the visual target stimuli caused an attenuation of the occipital N1 in the ERPs to the visual targets. As cued visual stimuli elicit enhanced N1 responses due to enhanced attention to these stimuli (Mangun & Hillyard, 1991), the attenuation of the N1 to the cued visual targets caused by a preceding deviant tone indicates that the involuntary switching of the subject's attention to the deviant tone interfered with the early, attentive visual processing (Alho et al., 1997).

ERPs recorded by Escera et al. (1998) showed that a deviant tone in the auditory-visual stimulus pair elicited MMN followed by a small P3a (see Figure 1). These ERP results, in combination with the behavioral results discussed above, suggest that two different attention-switching mechanisms are involved in the behavioral distraction observed in the auditory-visual distraction paradigm. The N1 enhancement to novel sounds, in comparison with the N1 to standard tones, was probably caused by an enhancement of the supratemporal N1 component (Näätänen & Picton, 1987) or some other N1 component related to attention, such as the frontal N1 described by Giard et al. (1994). However, MMN also appeared to contribute to the enhanced N1 deflection to the novel sounds occurring among standard tones, as recently demonstrated by Alho et al. (1998). Consequently, the switching of attention to novel sounds was probably triggered by a combined response of the transient-detector mechanism reflected by N1 and the stimulus-change detector mechanism reflected by MMN (cf. Näätänen, 1990, 1992; Näätänen & Picton, 1987). This process resulted in effective orienting of attention towards the novel sounds, as indicated by the subsequent large P3a wave illustrated in Figure 1 and the markedly delayed RT to the following visual stimulus illustrated in Figure 4. For deviant tones, the MMN and the subsequent small P3a suggest an attention switch initiated by the stimulus-change detector mechanism generating the MMN (Schröger, 1996). Apparently, this mechanism is less effective in triggering attention switches, as indicated by the smaller P3a, and the more modest effect on the visual RT compared with that caused by the novel sounds.

In the study of Escera et al. (1998), P3a response to novel sounds had two distinct subcomponents. Figure 5 illustrates that the early portion of P3a showed a centrally dominant scalp distribution with a polarity reversal at posterior and inferior lateral electrode sites, whereas the late portion of P3a displayed a frontal scalp maximum. The scalp distributions of these two P3a phases are in agreement with studies indicating multiple generator sources for P3a. The earlier, centrally maximal portion of P3a is probably dominated by contributions from posterior sources in the temporal and parietal cortices, indicated by effects of local brain lesions on ERPs (Knight et al., 1989), by intracranial ERP recordings (Halgren et al., 1995a, 1995b), and by ERP and MEG source modeling (Alho et al., 1998; Mecklinger & Ullsperger, 1995). Additionally, the later frontally maximal portion of P3a is presumably dominated by activity generated in anterior, prefrontal sources indicated by lesion studies (Knight, 1984), intracranial recordings (Baudena et al, 1995), and source modeling (Mecklinger & Ullsperger, 1995). It is likely that the two phases of P3a reflect two different processes in the course of involuntary attention switching, as also suggested by their different sensitivity to attentional manipulations. For example, Escera et al. (1998) found that the early P3a to novel sounds was of similar amplitude when their subjects concentrated on reading a book and when they performed the visual task in the distraction paradigm with auditory-visual stimulus pairs. The late P3a was enhanced in amplitude in the latter condition, in which the task-irrelevant auditory stimuli were to some extent covertly attended, as discussed above. The neural generators of the early phase of P3a and its apparent insensitivity to attentional manipulation suggest the early P3a partly reflects further processing of stimulus changes in the auditory cortex (Alho et al., 1998) and some violation of a multimodal model of the external world maintained in the temporal-parietal association cortex (Yamaguchi & Knight, 1991). The late P3a, in turn, may be more closely related to the actual orienting of attention, as suggested by its dependence on attention and by its prefrontal origin.

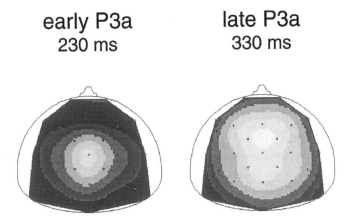

Figure 5. Voltage distribution maps for the P3a component elicited by infrequent, complex novel sounds occurring in a sequence of repetitive standard tones and infrequent, slightly higher deviant tones presented to a subject concentrating on a visual task. P3a amplitudes were measured at 230 and 330 milliseconds after sound onset from difference waves obtained by subtracting the event-related potentials (ERP) to standard tones from those to deviant tones. Lighter shades of gray indicate more positive voltages. As seen from these maps, the later phase of P3a is more frontally distributed than the earlier phase. For other details, see Figure 1 (after Yago et al., submitted).

The ERP data of Escera et al. (1998) also indicate that the P3a from novel sounds was followed by RON response, as shown in Figure 1, which appears to be associated with reorienting of attention back to the current task after a distracting event (Berti & Schröger, 2001; Schröger & Berti, 2000; Schröger et al., 2000; Schröger & Wolff, 1998a, b). Escera et al. (2001) recently provided evidence confirming the association of RON with reorienting of attention. They used a paradigm similar to that of Escera et al. (1998) except that a task-irrelevant auditory stimulus (that was either a standard tone, deviant tone or novel sound) occurred in different conditions that were either 245 or 355 milliseconds before each visual target stimulus. It was found that a negativity, presumably RON, following P3a responses to the deviant tones and novel sounds peaked at 350 milliseconds from visual-target onset in the two conditions of different auditory-visual stimulus onset asynchrony. Thus, RON was evidently related to the processing of visual target stimuli rather than to the processing of preceding task-irrelevant sounds.

3.   CONCLUSION

It has been long assumed that the brain activity reflected by MMN and P3a in ERPs to deviant sounds among repetitive standard sounds is associated with involuntary orienting of attention to auditory stimulus changes (Näätänen & Michie, 1979; Squires et al., 1975). However, the recent studies reviewed above measuring ERPs to infrequent changes in auditory stimuli and distracting effects of these changes on auditory and visual task performance have provided support for these early assumptions. These studies have also shown that involuntary attention to widely deviant auditory stimuli, e.g., novel sounds among tones, may be partly triggered by an enhanced N1 response to these sounds. Although MMN and N1 appear to indicate preattentive detection of a deviant or novel auditory event and initiation of an involuntary attention switch to such events, P3a is presumably associated with the actual consequent attention switching. Moreover, recent studies also suggest that reorienting of attention back to a task after a distracting stimulus change is reflected by RON, an ERP component following P3a after a distracting event.

Converging evidence from ERP and MEG studies using source-localization methods, as well as SCD analysis, intracranial ERP recordings, and studies examining effects of local brain lesions on ERPs in conjunction with studies applying PET, fMRI and EROS methods, indicate involvement of auditory and prefrontal cortices in the preattentive detection of auditory stimulus changes generating MMN and initiating an involuntary attention switch to these changes. According to MEG recordings, the enhancement of the N1 response to widely deviant sounds would also be explained by enhanced auditory-cortex activity, which might have a role, together with MMN, in initiation of involuntary attention to such sounds. Source localization of P3a, as well as intracranial P3a recordings and effects of brain lesions on P3a, in turn, suggests that a complex cerebral network, including areas in the prefrontal, temporo-parietal, and auditory cortices, as well as in the parahippocampal and anterior cingulate gyri and hippocampus, is activated during the actual attention switching to a deviant or novel auditory event. Finally, SCD analysis of RON implies that prefrontal areas are also involved in directing attention back to the task after a stimulus change distracts the task performance.

In addition to clarifying brain functions involved in involuntary attention to auditory stimuli, the experimental distraction paradigms described above have proven to be useful in studies on acute effects of alcohol on attention. Jääskeläinen et al. (1999) used the distraction paradigm developed by Schröger and Wolff (1998a) and found that both the RT prolongation and the P3a response caused by occasional task-irrelevant frequency changes in tones during a tone-duration discrimination task were reduced by a moderate dose of ethanol leading to about 0.04% blood alcohol concentration (BAC) relative to a placebo condition. Furthermore, Jääskeläinen et al. (1996) used the distraction paradigm developed by Escera et al. (1998) and found that the hit-rate reduction observed for visual target stimuli preceded by a deviant tone, as seen in Figure 4, was significantly smaller during mild ethanol intoxication (0.05% BAC) than in a placebo condition. Thus, even very small doses of ethanol that are generally regarded as not markedly deteriorating sensory-motor performance, e.g., driving a motor vehicle, suppress significantly the attention-capturing effects of changes in the auditory environment.

The distraction paradigms described above also have been used to determine impairment of attention in clinical groups. Ahveninen et al. (2000), using the paradigm developed by Schröger and Wolff (1998a), found that the RT prolongation in a tone-duration discrimination task caused by task-irrelevant changes in tone frequency was significantly larger in chronic alcoholics than in the control subjects. This RT effect correlated positively with the MMN amplitude (r=0.7). Hence, the abnormal attentional reactivity to irrelevant sound changes observed in the alcoholics appeared to be caused by over-active preattentive change detection generating MMN. Polo et al. (1998, 1999), also applied the auditory-visual distraction paradigm developed by Escera et al. (1998) and observed enhanced P3a responses to both frequency-deviant tones and novel sounds in chronic alcoholics in relation to control subjects, indicating enhanced involuntary orienting auditory stimulus changes in the alcoholic patients.

Finally, Kaipio et al. (in preparation) used the paradigm of Escera et al. (1998) to study closed-head injury patients with signs of distractibility during neuropsychological assessment. In these patients, the small auditory stimulus changes of the deviant tones were less effective in capturing attention compared to the healthy control subjects, as indicated by the absence of RT effects and RON after the deviant tones in the patients, whereas the controls showed similar RT distraction effects and RON after the deviant tones as in the study of Escera et al. (1998). The preattentive auditory deviance-detection system of the patients, however, worked quite normally, as indicated by approximately similar MMN and P3a responses in the patients and controls. The absence of RT effects and RON after deviant tones in the patients suggests that the patients were able to pay less attention to the auditory-visual stimulus relation than the controls. However, large auditory stimulus changes (novel sounds) were associated with an RT distraction effect and RON indicating that the attention-catching novel sounds caused a normal involuntary attention switch.

In conclusion, in addition to revealing the sequence of short-lived brain activations involved in involuntary attention and leading to distracted task performance, simultaneous measurements of ERPs and behavioral distraction caused by deviant sounds may provide valuable information on effects of ethanol and brain injuries on attention. Thus, combining behavioral and ERP measures appears to be a useful approach in studies on function and dysfunction of brain mechanisms involved in involuntary orienting of attention to changes in the auditory environment and in reorienting attention back to the distracted task.

ACKNOWLEDGMENTS

Supported by the Academy of Finland (grant number 49126), Generalitat de Catalunya (2001XT-00036), Spanish Ministry of Science and Technology (PM99-0167), DAAD, and Deutsche Forschungsgemeinschaft.

REFERENCES

Aaltonen, O., Tuomainen, J., Laine, M., & Niemi, P. (1993). Cortical differences in tonal versus vowel processing as revealed by an ERP component called mismatch negativity (MMN). *Brain and Language, 44,* 139-152.

Alho, K., Pesonaen, H., Keltikangas-Järvinen, L., Ravaja, N., Escera, C., Winkler, I., & Näätänen, R. (1999). Involuntary processing of novel sounds in individuals with high and low scores on the novelty-seeking temperament scale. *Psychophysiology, 36,* S24.

Alho, K., Winkler, I., Escera, C., Huotilainen, M., Virtanen, J., Jääskeläinen, I.P., Pekkonen, E., & Ilmoniemi, R.J. (1998). Processing of novel sounds and frequency changes in the human auditory cortex: Magnetoencephalographic recordings. *Psychophysiology, 35,* 211-224.

Alho, K., Woods, D.L., Algazi, A., & Näätänen, R. (1992). Intermodal selective attention II: Effects of attentional load on processing of auditory and visual stimuli in central space. *Electroencephalography and Clinical Neurophysiology, 82,* 356-368.

Ahveninen, J., Jääskeläinen, I.P., Pekkonen, E., Hallberg, A., Hietanen, M., Näätänen, R., Schröger, E., & Sillanaukee, P. (2000). Increased distractibility by task-irrelevant sound changes in abstinent alcoholics. *Alcoholism: Clinical and Experimental Research, 24,* 1850-1854.

Alain, C., Richer, F., Achim, A., & Saint-Hilaire, J.M. (1989). Human intracerebral potentials associated with target, novel and omitted auditory stimuli. *Brain Topography, 1,* 237-245.

Alain, C., Woods, D.L., & Knight, R.T. (1998). A distributed cortical network for auditory sensory memory in humans. *Brain Research, 812,* 23-37.

Alho, K., Escera, C., Díaz, R., Yago, E., & Serra, J.M. (1997). Effects of involuntary auditory attention on visual task performance and brain activity. *Neuroreport, 8,* 3233-3237.

Alho, K., Pesonen, H., Keltikangas-Järvinen, L., Ravaja, N., Escera, C., Winkler, I., & Näätänen, R. (1999). Involuntary processing of novel sounds in individuals with high and low scores on the novelty-seeking temperament scale[Abstract]. *Psychophysiology, 36,* S24.

Baudena, P., Halgren, E., Heit, G., & Clarke, J.M. (1995). Intracerebral potentials to rare target and distractor auditory and visual stimuli. III Frontal cortex. *Electroencephalography and Clinical Neurophysiology, 94,* 251-264.

Berti, S., & Schröger, E. (2001). A comparison of auditory and visual distraction effects: Behavioral and event-related indices. *Cognitive Brain Research, 11,* 265-273.

Celsis, P., Boulanouar, K., Doyon, B., Ranjeva, J.P., Berry, I., Nespoulous, J.L., & Chollet, F. (1999). Differential fMRI responses in the left posterior superior temporal gyrus and left supramarginal gyrus to habituation and change detection in syllables and tones. *Neuroimage, 9,* 135-144.

Cowan, N., Winkler, I., Teder, W., & Näätänen R. (1993). Memory prerequisites of the mismatch negativity in the auditory event-related potential (ERP). *Journal of Experimental Psychology: Learning, Memory and Cognition, 19,* 909-921.

Csépe, V., Karmos, G., & Molnár, M. (1987). Evoked potential correlates of stimulus deviance during wakefulness and sleep in cat-animal model of mismatch negativity. *Electroencephalography and Clinical Neurophysiology, 66,* 571-578.

Deouell, L., Bentin, S., & Giard, M.-H. (1998). Mismatch negativity in dichotic listening: evidence for interhemispheric differences and multiple generators. *Psychophysiology, 35,* 355-365.

Donchin, E., & Coles, M.G.H. (1988). Is the P300 component a manifestation of context updating? *Behavioral and Brain Sciences, 11,* 357-374.

Escera, C., Alho, K., Winkler, I., & Näätänen R. (1998). Neural mechanisms of involuntary attention switching to novelty and change in the acoustic environment. *Journal of Cognitive Neuroscience, 10,* 590-604.

Escera, C., Yago, E., & Alho, K. (2001). Electrical responses reveal the temporal dynamics of brain events during involuntary attention switching. *European Journal of Neuroscience, 14,* 877-883.

Escera, C., Corral, M.J., & Yago, E. An electrophysiological and behavioral investigation of involuntary attention towards auditory frequency, duration and intensity changes. *Cognitive Brain Research,* in press.

Ford, J.M., Roth, W.T., & Kopell, B.S. (1976). Auditory evoked potentials to unpredictable shifts in pitch. *Psychophysiology, 13,* 32-39.

Fuster, J. (1989). *The prefrontal cortex: Anatomy, physiology, and neuropsychology of the frontal lobe.* New York: Raven.

Giard, M.H., Perrin, F., Echallier, J.F., Thévenet, M., Froment, J.C., & Pernier J. (1994). Dissociation of temporal and frontal components in the auditory N1 wave: A scalp current density and dipole model analysis. *Electroencephalography and Clinical Neurophysiology, 92,* 238-252.

Giard, M.H., Lavikainen, J., Reinikainen, K., Perrin, F., Bertrand, O., Thévenet, M., Pernier, J., & Näätänen R. (1995). Separate representation of stimulus frequency, intensity, and duration in auditory sensory memory. *Journal of Cognitive Neuroscience, 7,* 133-143.

Giard, M.H., Perrin, F., Pernier, J., & Bouchet P. (1990). Brain generators implicated in processing of auditory stimulus deviance: A topographic event-related potential study. *Psychophysiology, 27,* 627-640.

Halgren, E., Baudena, P., Clarke, J.M., Heit, G., Liegeois, C., Chauvel, P., & Musolino, A. (1995a). Intracerebral potentials to rare target and distractor auditory and visual stimuli: I. Superior temporal plane and parietal lobe. *Electroencephalography and Clinical Neurophysiology, 94,* 191-220.

Halgren, E., Baudena, P., Clarke, J.M., Heit, G., Marinkovic, K., Devaux, B., Vignal, J.P., & Biraben, A. (1995b). Intracerebral potentials to rare target and distractor auditory and visual stimuli. II. Medial, lateral, and posterior temporal lobe. *Electroencephalography and Clinical Neurophysiology, 94,* 229-250.

Hari, R., Hämäläinen, M., Ilmoniemi, R., Kaukoranta, E., Reinikainen, K., Salminen, J., Alho, K., Näätänen, R., & Sams, M. (1984). Responses of the primary auditory cortex to pitch changes in a sequence of tone pips: Neuromagnetic recordings in man. *Neuroscience Letters, 50,* 127-132.

Hari, R., Kaila, K., Katila, T., Tuomisto, T., & Varpula, T. (1982). Interstimulus interval dependence of the auditory vertex response and its magnetic counterpart: Implications for their neural generation. *Electroencephalography and Clinical Neurophysiology, 54,* 561-569.

Jääskeläinen, I.P., Alho, K., Escera, C., Winkler, I., Sillanaukee, P., & Näätänen, R. (1996). Effects of ethanol and auditory distraction on forced choice reaction time. *Alcohol, 13,* 153-156.

Jääskeläinen, I.P., Schröger, E., & Näätänen, R. (1999). Effects of ethanol on auditory distraction: An ERP and behavioral study. *Psychopharmacology, 141,* 16-21.

Javitt, D.C., Steinschneider, M., Schroeder, C.E., & Arezzo, J.C. (1996). Role of cortical N-methyl-D-aspartate receptors in auditory sensory memory and mismatch negativity generation: Implications for schizophrenia. *Proceedings of the National Academy of Sciences USA, 93,* 11962-11967.

Kaipio, M.L., Escera, C., Winkler, I., Surma-aho, O., & Alho, K. Attention impairment in closed-head injury as revealed by event-related potentials. In preparation.

Knight, R.T. (1984). Decreased response to novel stimuli after prefrontal lesion in man. *Electroencephalography and Clinical Neurophysiology, 59,* 9-20.

Knight, R.T. (1991). Evoked potential studies of attention capacity in human frontal lobe lesions. In H. Levin, H., H. Eisenberg, & F. Benton (Eds.), *Frontal lobe function and dysfunction* (pp. 139-153). Oxford: Oxford University Press.

Knight, R.T. (1996). Contribution of human hippocampal region to novelty detection. *Nature, 383,* 256-259.

Knight, R.T., Scabini, D., Woods, D.L., & Clayworth, C. (1989). Contributions of temporal-parietal junction to the human auditory P3. *Brain Research, 502,* 109-116.

Knight, R.T., & Scabini, D. (1998). Anatomic bases of event-related potentials and their relationship to novelty detection in humans. *Journal of Clinical Neurophysiology, 15,* 3-13.

Korzyukov, O., Alho, K., Kujala, A., Gumenyuk, V., Ilmoniemi, R.J., Virtanen, J., Kropotov, J., & Näätänen, R. (1999). Electromagnetic responses of the human auditory cortex generated by sensory-memory based processing of tone-frequency changes. *Neuroscience Letters, 276,* 169-172.

Kraus, N., McGee, T., Carrell, T., King, C., Littman, T., & Nicol, T. (1994). Discrimination of speech-like contrasts in the auditory thalamus and cortex. *Journal of the Acoustical Society of America, 96,* 2758-2768.

Kropotov, J.D., Näätänen, R., Sevostianov, A.V., Alho, K., Reinikainen, K., & Kropotova, O.V. (1995). Mismatch negativity to auditory stimulus change recorded directly from the human temporal cortex. *Psychophysiology, 32,* 418-422.

Kropotov, J.D., Alho, K., Näätänen, R., Ponomarev, V.A., Kropotova, O.V., Anichkov, A.D., & Nechaev, V.B. (2000). Human auditory-cortex mechanisms of preattentive sound discrimination. *Neuroscience Letters, 280,* 87-90

Levänen, S., Ahonen, A., Hari, R., McEvoy, L., & Sams, M. (1996). Deviant auditory stimuli activate human left and right auditory cortex differently. *Cerebral Cortex, 6,* 288-296.

Liasis, A., Towell, A., & Boyd, S. (1999). Intracranial auditory detection and discrimination potentials as substrates of echoic memory in children. *Cognitive Brain Research, 7,* 503-506.

Lyytinen, H., Blomberg, A.P., & Näätänen, R. (1992). Event-related potentials and autonomic responses to a change in unattended auditory stimuli. *Psychophysiology, 29,* 523-534.

Mangun, G.R., & Hillyard, S.A. (1991). Modulations of sensory-evoked brain potentials indicate changes in perceptual processing during visual-spatial priming. *Journal of Experimental Psychology: Human Perception and Performance, 17,* 1057-1074.

Mecklinger A., & Ullsperger P. (1995). The P300 to novel and target events: A spatio-temporal dipole model analysis. *Neuroreport, 7,* 241-245.

Näätänen, R. (1990). The role of attention in auditory information processing as revealed by event-related potentials and other brain measures of cognitive function. *Behavioral and Brain Sciences, 13,* 201-288.

Näätänen, R. (1992). *Attention and brain function.* Hillsdale, NJ: Lawrence Erlbaum Associates.

Näätänen, R. (1995). The mismatch negativity: A powerful tool for cognitive neuroscience. *Ear and Hearing, 16,* 6-18.

Näätänen, R., & Alho K. (1997). Mismatch negativity (MMN): The measure for central sound representation accuracy. *Audiology and Neuro-Otology, 2,* 341-353.

Näätänen, R., Gaillard, A.W.K., & Mäntysalo, S. (1978). Early selective attention effect on evoked potential reinterpreted. *Acta Psychologica, 42,* 313-329.

Näätänen, R., Gaillard, A.W.K., & Mäntysalo, S. (1980). Brain potential correlates of voluntary and involuntary attention. In H. H. Kornhuber, & L. Deecke (Eds.), *Motivation, motor and sensory processes of the brain: Electrical potentials, behavior and clinical use. Progress in brain research, Vol. 54* (pp. 343-348). Amsterdam: Elsevier.

Näätänen, R., & Michie, P.T. (1979). Early selective attention effects on the evoked potential. A critical review and reinterpretation. *Biological Psychology, 8*, 81-136.

Näätänen, R., Paavilainen, P., Alho, K., Reinikainen, K., & Sams, M. (1989). Do event-related potentials reveal the mechanism of the auditory sensory memory in the human brain? *Neuroscience Letters, 98*, 217-221.

Näätänen, R., Paavilainen, P., Tiitinen, H., Jiang, D., & Alho, K. (1993). Attention and mismatch negativity. *Psychophysiology, 30*, 436-450.

Näätänen, R., & Picton, TW. (1987). The N1 wave of the human electric and magnetic response to sound: A review and an analysis of the component structure. *Psychophysiology, 24*, 375-425.

Näätänen, R., & Winkler, I. (1999). The concept of auditory stimulus representation in neuroscience. *Psychological Bulletin, 125*, 826-859.

Öhman, A. (1979). The orienting response, attention, and learning: An information-processing perspective. In H.D. Kimmel, H.E. van Olst, & J.F. Orlebeke, J.F. (Eds.), *The orienting response in humans* (pp. 443–471). Hillsdale, NJ: Lawrence Erlbaum Associates.

Opitz, B., Mecklinger, A., Von Cramon, D.Y., & Kruggel, F. (1999). Combined electrophysiological and hemodynamic measures of the auditory oddball. *Psychophysiology, 36*, 142-147.

Opitz, B., Rinne, T., Mecklinger, A., von Cramon, D.Y., & Schröger, E. (2002). Differential contribution of frontal and temporal cortices to auditory change detection: fMRI and ERP results. *NeuroImage, 15*, 165-174.

Polo, M.D., Yago, E., Grau, C., Alho, K., Gual, A., & Escera, C. (1998). Involuntary attentional processing in chronic alcoholism. In M. Tervaniemi, & C. Escera (Eds.), *Abstracts of the first international workshop on mismatch negativity and its clinical applications* (p. 88). Helsinki: Multiprint-University of Helsinki.

Polo, M.D., Yago, E., Gual, A., Grau, C., Alho, K., & Escera, C. (1999). Abnormal activation of cerebral networks of orienting to novelty in chronic alcoholics. *Psychophysiology, 36*, S90.

Rinne, T., Alho, K., Alku, P., Holi, M., Sinkkonen, J., Virtanen, J., Bertrand, O., & Näätänen, R. (1999a). Analysis of speech sounds is left-hemisphere predominant at 100-150 ms after sound onset. *Neuroreport, 10*, 1113-1117.

Rinne, T., Alho, K., Ilmoniemi, R.J., Virtanen, J., & Näätänen, R. (2000). Separate time behaviors of the temporal and frontal mismatch negativity sources. *Neuroimage, 12*, 14-19.

Rinne, T., Gratton, G., Fabiani, M., Cowan, N., Maclin, E., Stinard, A., Sinkkonen, J., Alho, K., & Näätänen, R. (1999). Scalp-recorded optical signals make sound processing in the auditory cortex visible. *Neuroimage, 10*, 620-624.

Ritter, W., Deacon, D., Gomes, H., Javitt, D.C., & Vaughan, H.G., Jr. (1995). The mismatch negativity of event-related potentials as a probe of transient auditory memory: A review. *Ear and Hearing, 16*, 52-67.

Sams, M., Paavilainen, P., Alho, K., & Näätänen, R. (1985). Auditory frequency discrimination and event-related potentials. *Electroencephalography and Clinical Neurophysiology, 62*, 437-448.

Scherg, M., Vajsar, J., & Picton, T. (1989). A source analysis of the human auditory evoked potentials. *Journal of Cognitive Neuroscience, 1*, 336-355.

Schröger, E. (1996). A neural mechanism for involuntary attention shifts to changes in auditory stimulation. *Journal of Cognitive Neuroscience, 8*, 527-539.

Schröger, E. (1997). On the detection of auditory deviants: A pre-attentive activation model. *Psychophysiology, 34*, 245-257.

Schröger, E., & Berti, S. (2000). Distracting working memory by automatic deviance-detection in audition and vision. In E. Schröger, A. Mecklinger, & A.D. Friederici (Eds.), *Working on working memory* (pp. 1-25). Leipzig: Leipzig University Press.

Schröger, E., Giard, M.H., & Wolff, C. (2000). Auditory distraction: Event-related potential and behavioral indices of auditory distraction. *Clinical Neurophysiology, 111*, 1450-1460.

Schröger, E., & Wolff, C. (1998a). Attentional orienting and reorienting is indicated by human event-related brain potentials. *Neuroreport, 9*, 3355-3358.

Schröger, E., & Wolff, C. (1998b). Behavioral and electrophysiological effects of task-irrelevant sound change: A new distraction paradigm. *Cognitive Brain Research, 7*, 71-87.

Schröger, E., & Wolff, C. Auditory distraction as a function of the discrepancy of the distractor. In preparation.

Serra, J.M., Giard, M.H., Yago, E., Alho, K., & Escera, C. (1998). Bilateral contribution from frontal lobes to MMN. *International Journal of Psychophysiology, 30,* 236-237.

Sokolov, E.N. (1975). The neuronal mechanisms of the orienting reflex. In E.N. Sokolov & O.S. Vinogradova (Eds.), *Neuronal mechanisms of the orienting reflex* (pp. 217–338). Hillsdale, NJ: John Wiley.

Squires, K.C., Squires, N.K., & Hillyard, S.A. (1975). Decision-related cortical potentials during an auditory signal detection task with cued observation intervals. *Journal of Experimental Psychology: Human Perception and Performance, 1,* 268-279.

Stuss, D.T., & Benson, D.F. (1986). *The frontal lobes.* New York: Raven Press.

Sutton, S., Braren, M., Zubin, J., & John, E.R. (1965). Evoked-potential correlates of stimulus uncertainty. *Science, 150,* 1187-1188.

Tervaniemi, M., Medvedev, S.V., Alho, K., Pakhomov, S.V., Roudas, M.S., van Zuijen, T.L., & Näätänen, R. (2000). Lateralized automatic auditory processing of phonetic versus musical information: A PET study. *Human Brain Mapping, 10,* 74-79.

Tiitinen, H., May, P., Reinikainen, K., & Näätänen, R. (1994). Attentive novelty detection in humans is governed by pre-attentive sensory memory. *Nature, 372,* 90-92.

Winkler, I., & Czigler, I. (1998). Mismatch negativity: Deviance detector or the maintenance of the "standard." *Neuroreport, 9,* 3809-3813.

Winkler, I., Karmos, G., & Näätänen, R. (1996), Adaptive modeling of the unattended acoustic environment reflected in the mismatch negativity event-related potential. *Brain Research, 742,* 239-252.

Woods, D.L. (1990). The physiological basis of selective attention: Implications of event-related potential studies. In J.W. Rohrbaugh, R. Parasuraman, & R. Johnson, Jr. (Eds.), *Event-related potentials: Basic issues and applications* (pp. 178-209). New York: Oxford University Press.

Woods, D.L. (1992). Auditory selective attention in middle-aged and elderly subjects: An event-related brain potential study. *Electroencephalography and Clinical Neurophysiology, 84,* 456-468.

Woods, D.L., Knight, R.T., & Scabini, D. (1993). Anatomical substrates of auditory selective attention: Behavioral and electrophysiological effects of posterior association cortex lesions. *Cognitive Brain Research, 1,* 227-240.

Yamaguchi, S., & Knight, R.T. (1991). Anterior and posterior association cortex contributions to the somatosensory P300. *Journal of Neuroscience, 11,* 2039-2054.

Yago, E., Escera, C., Alho, K., & Giard, M.H. (2001). Cerebral mechanisms underlying orienting of attention towards auditory frequency changes. *Neuroreport, 12,* 2583.2587.

Yago, E., Escera, C., Alho, K., Giard, M.H., & Serra-Grabulosa, J.M. Spatiotemporal dynamics of the auditory novelty-P3 event-related brain potential. *Cognitive Brain Research.* Submitted.

Chapter 3

VISUAL MISMATCH NEGATIVITY

DIRK J. HESLENFELD
Department of Psychology, Free University, Amsterdam, The Netherlands

1. INTRODUCTION

When a deviant tone is presented within a sequence of repetitive ("standard") tones, it evokes a specific response in the event-related potential (ERP) called mismatch negativity (MMN). This response lasts from about 100 to 250 milliseconds after stimulus onset, is maximal over frontal/central scalp areas, and is thought to originate from temporal and frontal cortices (for reviews, see Näätänen, 1990, 1992, 1995). Näätänen defines the MMN as "the brain's automatic response to changes in repetitive auditory input" (1990, p. 201). Within the auditory modality, MMN has been observed to changes in tonal frequency, intensity, duration, spatial location, and many other auditory stimuli parameters. An important part of its definition is the automaticity of the response—its independence of attention. Indeed, it has been shown many times that the auditory MMN is unaffected to a large degree by the difficulty (or load) of a concurrent task in the visual modality (Alho et al., 1992; Sams et al., 1985, Ritter et al., 1995). Note that this independence of attention does not imply that the auditory process underlying MMN occur without any attention. An alternative possibility is that the auditory process has its own, modality-specific attentional resource (Wickens, 1984), in which case MMN will be unaffected by the difficulty of a task in a different modality. Indeed, effective withdrawal of auditory attention by a concurrent task in the auditory modality has been shown to reduce the MMN (Alain & Woods, 1997; Woldorff et al., 1991, 1998).

It is still unclear whether a comparable MMN component can be obtained in the visual modality. Nyman et al. (1990) presented sine wave gratings at a rate of one in 490 milliseconds to the central (2°×2°) visual field. The standard gratings (90%) had a high spatial contrast of 0.72, whereas the deviant gratings (10%) had a lower spatial contrast of 0.24. Between 200 and 300 milliseconds after stimulus onset, ERPs elicited by the low contrast deviant stimuli were negatively displaced with respect to ERPs elicited by high contrast standard stimuli. However, when the mapping of contrasts on stimulus probabilities was reversed, the negative displacement occurred to all low contrast stimuli irrespective of their probability. Hence, the negative displacement was not due to the deviance of the contrast but rather to the low contrast as such—that is, it was an exogenous effect. However, the negative displacement in the control experiment occurred between 100 and 200 milliseconds, which leaves the 200-300 millisecond effect in the main experiment unexplained.

Czigler and Csibra (1990) presented a small (76'×46') black rectangle once in 417 milliseconds to the central visual field. On 10% of the trials, the rectangle was slightly thicker than on the remaining 90% of the trials. The subjects did not notice the difference between standard and deviant thickness, and the ERP waveforms did not differ from each other. After the subjects had been notified of the difference, a small negative response (210-240 milliseconds) to deviant stimuli was observed in a second block. In addition, a deviant orientation of two easily visible angles inside the rectangle always led to negative deflections between 90 and 180 milliseconds and between 210 and 270 milliseconds. Whereas the responses after 200 milliseconds are likely originated from attention-related processes (Harter & Guido, 1980), the earlier 90-180 millisecond negativity might be a visual analogue of the auditory MMN (see also Czigler & Csibra, 1992). No control for possible exogenous effects of line orientation was imposed, so that the observed early difference might also be due to the physical difference between standard and deviant stimuli (Harter et al., 1980).

Woods et al. (1992) presented auditory and visual stimuli at a mean rate of one in 300 milliseconds. The visual stimuli were vertical gratings of either 0.7 or 2.0 cycles per degree (cpd) and were flashed either to the left or right of fixation (at 4.7° eccentricity). The standard stimuli 90% were slightly taller than wide (3.9°×4.4°), the remaining 10% were shorter (about 3.9°×3.6°). In different blocks, either visual or auditory deviants required a response. A negative deflection was observed to stimuli with the deviant size (or shape) at contralateral posterior scalp sites from about 160 to 360 milliseconds. This negativity occurred in both the auditory and visual attention conditions, but was 3.3 times larger during visual (intramodal) attention. Although this response may be a visual MMN, it may also exogenously reflect the constant difference in size between standard and deviant stimuli (Harter et al., 1976).

In a companion paper, Alho et al. (1992) flashed the 2.0 cycles per degree (cpd) gratings to the central visual field: standard stimuli (80%) were again taller than wide (3.9°×4.4°), 10% were slightly shorter (3.9°×3.9°) small deviants, and 10% were considerably shorter (3.9°×2.4°) large deviants. In different blocks, either small or large deviants in either the auditory or visual modality required a response. The small visual deviants elicited a negativity only when they were targets and only after 270 milliseconds. The large visual deviants elicited negativities at 120, 200, and 270 milliseconds, which were enhanced when the visual stimuli were attended. The first response at 120 milliseconds, however, was unaffected by intermodal attention, which may point to an automatic processing of the deviant visual feature. Again, however, the deviant stimuli always had a different size than the standard stimuli, so that these effects might be due to the different stimulus parameters. A control experiment presented the large deviants alone (i.e., without intervening standards), and found the same negative responses as when the deviants were embedded within standards (Alho et al., 1992). This result strongly suggests an exogenous cause of the observed effect.

Tales et al. (1999) presented either one thick bar (2.2°× .68°) or two thinner bars (2.2°×.34°) every 612-642 milliseconds to the upper and lower part of a computer screen. Subjects' attention was directed to the center of the screen by asking them to detect an occasional change of a color patch. In experiment 1, the thick bars were presented frequently (88.9%) and the thinner bars infrequently (5.6%); in experiment 2, this mapping was reversed. They found a late and long-lasting negativity (250-400 milliseconds) at all posterior electrodes to deviant stimuli that was independent of which stimulus served as deviant. There was no control over the attention of the subjects, other than that they had to detect an occasional (5.6%) change of a color patch at the center of the screen. Moreover, the long latency and broad scalp distribution of the effect highly resembles the classical visual selection negativity (Harter & Guido, 1980; for a review, see Heslenfeld et al., 1997). Thus, this result more likely reflects attentive processing of deviant stimuli rather than an early, automatic, visual mismatch negativity (cf. Czigler & Csibra, 1990).

In sum, previous studies of visual deviance-related processes either have observed no effect at all, or have failed to control for confounding exogenous stimulus parameters. In addition, the stimulus deviance dimension in previous studies was always relevant to the task during some blocks (i.e., during some visual attention conditions), so that it cannot be excluded that stimuli were differentially processed in this respect even during auditory attention conditions. In the present experiment, exogenous factors are controlled by comparing ERPs to physically identical stimuli that were either standards or deviants in different blocks. None of these stimuli were ever task-relevant, excluding any residual task-related processing by deviant stimuli. Finally, three levels of difficulty of a concurrent task within the same visual modality were employed to assess the automaticity of possible deviance-related effects.

2.   METHODS

2.1   Subjects

Fourteen subjects (age range 18-24, mean age 20.9 years, 10 females, 12 right-handed) participated for course credits. All had normal or corrected-to-normal vision; none reported any history of neurological or psychiatric disease.

2.2   Stimuli, Task, and Procedure

The only task of the subjects was a compensatory visuo-motor tracking task. A small bright rectangle was continuously visible at the center of the computer screen, which moved constantly and unpredictably either to the left or right. The task of the subject was to keep the rectangle in the middle of the screen by means of compensatory button presses with the left and right index finger. The speed of rectangle movement and the frequency of its spontaneous changes of direction were linearly adjusted in order to create three levels of task difficulty. Relative to the "easy" condition, the speed and frequency of direction changes were doubled in the "moderate" condition, and tripled in the "difficult" condition. A total of 24 blocks each lasting 3.5 minutes were presented to each subject, whose only task during the entire experiment was to keep the moving rectangle in the middle of the screen.

Two types of task-irrelevant probe stimuli were presented: (a) Every 4-14 seconds, the screen became blank for 33.3 milliseconds, which appeared as a brief white flash. The data to these flashes will not be discussed. (b) In addition, white-on-black vertical square wave gratings were presented simultaneously to the upper and lower $5.6°$ of the computer screen in 12 of the 24 blocks. A central horizontal bar 45' in height was not stimulated to avoid interference with the horizontally moving rectangle. This area was constantly demarcated by two thin (1.5') white horizontal lines. Two small rectangles (4.5'×7.5') were attached to these lines to continuously indicate the mid-screen target position.

The gratings had either a high (2.3 cpd) or low (0.58 cpd) fundamental spatial frequency, and they were presented for 16.7 milliseconds once in 350-450 milliseconds (rectangular distribution), with a contrast of about 20%. In 6 of the 12 blocks, the low spatial frequency grating was presented in 80% of the trials, and the high spatial frequency grating in the remaining 20%. In the other 6 blocks, the presentation rates were reversed. A total of 400 trials with the standard spatial frequency and 100 trials with the deviant spatial frequency were randomly intermixed in each block, with the exception that a deviant grating was always preceded by a standard grating. Task difficulty was constant for four blocks in a row, after which it was changed to another level. Two of these four blocks contained only flashes, the other two contained flashes and gratings. Blocks with only flashes and blocks with flashes and gratings were alternated. After 12 blocks (4 blocks at each difficulty level), the entire sequence of events was replicated with another 12 blocks. The sequences of task difficulties, whether or not there were gratings in a block, and which grating was the standard, were pseudo-randomized and counterbalanced over subjects. Subjects sat in a silent, dimly illuminated room, and were trained on all task difficulties before the experiment began.

2.3 Recordings

Electroencephalographic (EEG) data were recorded from eight tin electrodes mounted in an elastic cap (at locations Fz, Cz, Pz, Oz, C3, C4, T5, T6), and one electrode on the left mastoid all referenced to an electrode on the right mastoid, with impedances kept below 5kΩ. Vertical electro-oculogram (EOG) activity was recorded bipolarly from above and below the right eye, and horizontal EOG from the outer canthi of each eye. Data acquisition was continuous, with a sampling rate of 250 Hz and band pass filtering at 0.08-35 Hz. The left and right button presses (binary responses) and the position of the rectangle on the screen (in pixels) were also recorded.

2.4 Data analysis

2.4.1 Performance

The root mean square of the lateral position of the moving rectangle was computed to estimate the performance during each block. That is, deviations of the rectangle from the mid-screen target position were squared and averaged over time, separately for each block. The square roots of these means were averaged across replications. The performance data were analyzed by a Task Load (easy, moderate, difficult) × Standard Spatial Frequency (high, low) multivariate analysis of variance (MANOVA). Hotelling's T^2 test was used to assess effects involving Task Load.

2.4.2   Evoked Potentials

Time series of 512 milliseconds duration were extracted off-line from the continuous data for each of the 14 channels, time-locked to the onset of a grating. The first 60 milliseconds prior to the onset of the grating defined the pre-stimulus baseline. All EEG channels were re-referenced to the algebraic mean amplitudes of the two mastoids. To reduce the low frequency drifts that might occur if subjects followed the moving rectangle with their eyes, linear trends were fitted and removed from all raw EEG and EOG time series (Porges & Bohrer, 1990). EOG artifacts were removed from the EEG by the method of Woestenburg et al. (1983). Trials containing amplifier saturations, peak-to-peak amplitude differences larger than 120 µV, or sample-to-sample amplitude differences larger than 25 µV, were discarded. In addition, the first 10 trials of each block were excluded from the averages.

Trials were averaged according to task difficulty, spatial frequency, and stimulus probability, and then were averaged across replications. For each EEG channel, 18 mean amplitude measures of 20 milliseconds duration (from 40 to 400 milliseconds post-stimulus) were computed, which reflected the mean of the recorded voltage in each condition for each scalp site and time interval. These data were analyzed by MANOVAs, with the factors Task Load (easy, moderate, difficult), Spatial Frequency (high, low), Deviance (standard, deviant), and Channels (3 levels). The analyzed channels were Fz, Cz, Pz (midline), and T5, Oz, T6 (posterior row, in a separate analysis). If necessary, scalp distributions were normalized such that their multivariate vector lengths equaled one (McCarthy & Wood, 1985). Hotelling's T^2 was used for all tests involving factors with more than 2 levels. Because of the large number of tests, the critical α level was set to 0.01. Since almost all interesting effects occurred at Oz between 60 and 100 milliseconds, and at Fz and Oz between 120 and 180 milliseconds, separate Task Load × Spatial Frequency × Deviance analyses were performed for these mean amplitude measures. The critical α level for these additional analyses was 0.05.

3.   RESULTS

3.1   Performance

Figure 1 displays the means and standard errors (across 14 subjects) of the performance data in the three task load conditions, separately for blocks in which high or low spatial frequency gratings were standards. Data are shown in pixels; 40 pixels correspond to 1° of visual angle. There was a main effect of Task Load ($F(2,12)=919.72$, $p<0.001$), and a main effect of Spatial Frequency ($F(1,13)=7.49$, $p<0.017$), but no interaction. A separate 2×2 analysis involving the "easy" and "moderate" conditions showed only a main effect of Task Load ($F(1,13)=1412.03$, $p<0.001$), but no effects of Spatial Frequency. A 2×2 analysis involving the "moderate" and "difficult" conditions revealed main effects of Task Load ($F(1,13)=184.33$, $p<0.001$) and Spatial Frequency ($F(1,13)=5.46$, $p<0.037$), but no interaction. Thus, performance in both the "easy" and the "difficult" conditions differed from the "moderate" condition, and in the more difficult conditions low spatial frequency standards interfered more with the task than high spatial frequency standards.

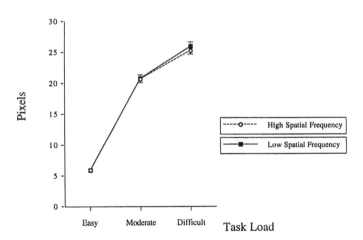

Figure 1. Performance data for each level of task difficulty for high and low standard spatial frequencies (task-irrelevant probes). Shown are means and standard errors of the root mean squared deviations of the moving rectangle from the target mid-screen position (n=14). Data are given in pixels; 40 pixels correspond to 1° of visual angle.

3.2 Evoked Potentials

Figure 2 displays the grand average (over 14 subjects) evoked potentials at eight electrodes to each grating at each probability level in the "moderate" task load condition. There was a large negative response to the high spatial frequency gratings at Oz between 50 and 100 milliseconds, which seemed to be larger for deviant than standard high spatial frequency gratings and appears to correspond to the early exogenous C1 component (Smith & Jeffreys, 1978). This response was followed at Oz by both spatial frequency and deviance effects from about 100 to 200 milliseconds. In addition, there are deviance effects at Fz and Cz in the same latency range, which seem larger for the low than the high spatial frequency gratings.

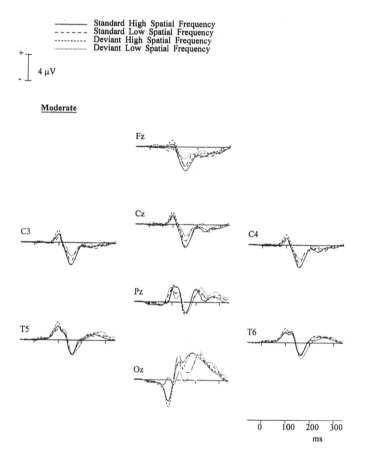

Figure 2. Grand averaged evoked potentials at 8 scalp sites to low and high, standard and deviant spatial frequencies from the intermediate level of task difficulty (n=14).

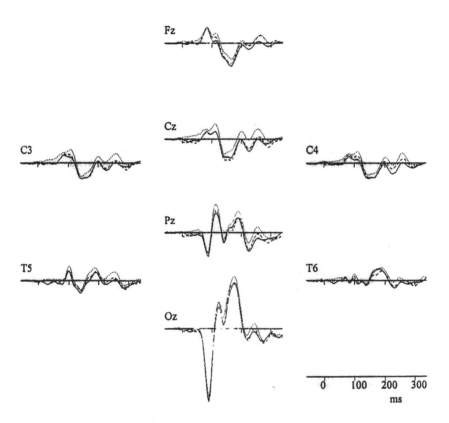

High - Low Spatial Frequency

Figure 3. Grand average differences between responses evoked by high and low spatial frequencies at three levels of task difficulty (n=14).

Figure 3 shows the differences between responses to high and low spatial frequencies, pooled over stimulus probabilities, separately for each task load condition. Starting with the posterior channels (T5, Oz, T6), there was an interaction between Spatial Frequency and Channels from 60 to 100 milliseconds (smallest $F(2,12)=22.93$, $p<0.001$), accompanied by a main effect of Spatial Frequency from 80 to 100 milliseconds ($F(1,13)=24.88$,

$p<0.001$), reflecting a larger negative response to high spatial frequency gratings, which was larger at Oz than at T5 or T6. There was a second interaction between Spatial Frequency and Channels, plus a Spatial Frequency main effect, between 160 and 200 milliseconds, reflecting a larger positive response to high spatial frequency gratings, which was again larger at Oz than at T5 and T6 (smallest $F(2,12)=11.06$, $p<0.002$, and smallest $F(1,13)=20.21$, $p<0.001$). There was no interaction between Spatial Frequency and Task Load, or Spatial Frequency, Task Load, and Channels. In other words, these exogenous effects were independent of task load, and thus replicable across blocks.

At the midline leads (Fz, Cz, Pz), there was a Spatial Frequency × Channels interaction from 60 to 120 milliseconds (smallest $F(2,12)=10.15$, $p<0.003$), which reflects the scalp distribution of the spatial frequency effect (posteriorly negative, anteriorly positive). The interaction was followed by a main effect of Spatial Frequency from 140 to 160 milliseconds (a negativity, $F(1,13)=13.09$, $p<0.004$), which was followed by another interaction with Channels from 180 to 200 milliseconds (posteriorly positive, anteriorly negative, $F(2,12)=7.14$, $p<0.01$). There was no interaction between Spatial Frequency and Task Load, or Spatial Frequency, Task Load, and Channels, again stressing the robustness of these exogenous effects to both task load and replication.

Figure 4 displays the differences between the responses to standard and deviant gratings, pooled over spatial frequencies, separately for each task load condition. At the posterior channels, there was a Deviance × Channels interaction from 60 to 160 milliseconds (smallest $F(2,12)=7.20$, $p<0.009$), which was overlapped and followed by a Deviance main effect from 120 to 200 milliseconds (smallest $F(1,13)=10.61$, $p<0.007$). This sequence of effects seemed due to the fact that (Figure 4) the deviance effect was negative at Oz and slightly positive at T5/T6 from 60 to 120 milliseconds, negative and larger at Oz than at T5/T6 from 120 to 160 milliseconds, and negative at all posterior electrodes from 160 to 200 milliseconds. There was no significant interaction between Deviance and Task Load, or Deviance, Task Load, and Channels, showing that these effects were independent of the difficulty of the visual tracking task. In addition, there was no interaction between Deviance and Spatial Frequency, but there was a significant Deviance × Spatial Frequency × Channels interaction between 80 and 100 milliseconds ($F(2,12)=17.46$, $p<0.001$), indicating that the earliest part of the deviance effect was a modulation of the scalp topographies of the C1 responses to the two spatial frequencies. Indeed, this effect vanished after normalizing the scalp distributions ($F(2,12)=0.63$, $p<0.55$), so that it could be attributed to an amplitude modulation of the C1 response in particular to high spatial frequency gratings by stimulus deviance (also see Figure 5).

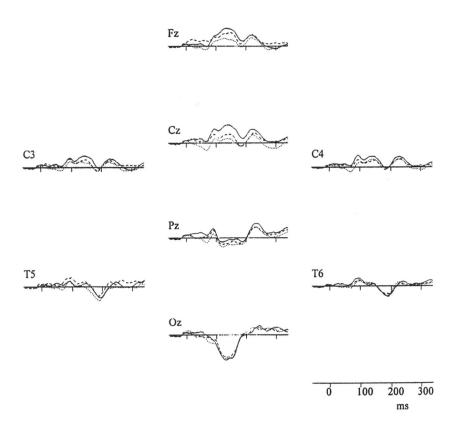

Deviant - Standard Spatial Frequency

Figure 4. Grand average differences between responses evoked by standard and deviant spatial frequencies at three levels of task difficulty (n=14).

At the midline leads, there was a Deviance × Channels interaction from 120 to 140 milliseconds ($F(2,12)=10.05$, $p<0.003$), a trend towards a Task Load × Deviance interaction from 120 to 180 milliseconds (smallest $F(2,12)=5.83$, $p<0.018$), and a Task Load × Deviance × Channels interaction from 140 to 180 milliseconds (smallest $F(4,10)=6.05$, $p<0.01$), which was preceded by a trend from 120 to 140 milliseconds ($F(4,10)=5.77$, $p<0.012$).

Thus, as illustrated in Figure 4, deviant stimuli evoked a positive response between 120 and 180 milliseconds, which was larger in the easier task conditions and larger at frontal than at parietal scalp sites.

To summarize the results so far, there was (1) a Spatial Frequency effect, largest at Oz from 60 to 100 milliseconds, which was independent of Task Load, but amplified by Deviance; (2) a second Spatial Frequency effect between about 140 and 200 milliseconds, which was independent of both Task Load and Deviance; (3) a posterior Deviance effect (120-200 milliseconds), which was independent of Spatial Frequency and Task Load; and (4) an anterior Deviance effect (120-180 milliseconds), which was larger for easier Task Load conditions. Thus, the posterior and anterior deviance effects were dissociated by their differential dependence on task load. In order to further clarify these effects, mean amplitudes between 60 and 100 milliseconds were computed at Oz, and between 120 and 180 milliseconds at Fz and Oz.

Deviance Effects: 60-100 ms

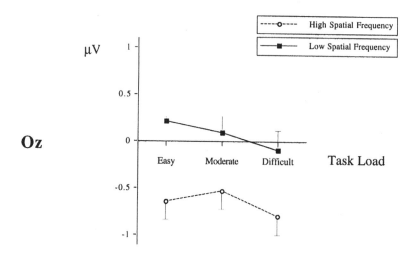

Figure 5. Means (+SEM) for the deviance effect at Oz between 60 and 100 milliseconds (n=14). These points reflect the difference between amplitudes to standard and deviant spatial frequencies, separately for each spatial frequency and task load condition.

Figure 5 shows the deviance effect (i.e., the *difference* between responses evoked by standard and deviant gratings) at Oz in the time interval from 60 to 100 milliseconds, separately for each task load and spatial frequency. A

Spatial Frequency × Deviance × Task Load MANOVA revealed a main effect of Deviance ($F(1,13)=5.86$, $p<0.031$) and a Spatial Frequency × Deviance interaction ($F(1,13)=11.60$, $p<0.005$), but no Task Load effects. The interaction indicated that, irrespective of task load, only responses to high spatial frequencies were affected by deviance: Their C1 response was significantly larger when they were deviant than when they were standard (mean difference -0.65 µV, $SE= 0.12$; $F(1,13)=16.96$, $p<0.002$). In contrast, the low spatial frequency response in the same time interval was not affected by stimulus deviance (mean difference $+0.06$ µV, $SE= 0.11$; $F(1,13)=0.18$, $p<0.68$). Although Figure 5 suggests a modulation by task load of the deviance effect for the low spatial frequencies, this effect failed to reach significance when tested separately ($F(2,12)=2.94$, $p<0.10$).

Figure 6 shows the same differences (i.e., deviant–standard spatial frequencies) for the time interval from 120 to 180 milliseconds at Fz (upper panel) and Oz (lower panel). At Oz there was a main effect of Deviance ($F(1,13)=84.88$, $p<0.001$), which did not depend on Task Load or Spatial Frequency. Thus, the magnitudes of the deviance effects were not significantly different for the two spatial frequencies (low spatial frequency: mean difference -1.33 µV, $SE= 0.16$; high spatial frequency: mean difference -0.86 µV, $SE= 0.12$; test on difference: $F(1,13)=2.22$, $p<0.16$).

At Fz, there was a main effect of Deviance ($F(1,13)=13.94$, $p<0.003$), and an interaction between Deviance and Task Load ($F(2,12)=6.49$, $p<0.013$). However, contrary to what is suggested by Figure 6, no effect involving Spatial Frequency was significant. However, since the deviance effect clearly seemed larger for low spatial frequencies in the "easy" task load condition, we ran separate Deviance × Task Load MANOVAs for each spatial frequency. For the high spatial frequency, there was a main effect of Deviance ($F(1,13)=8.15$, $p<0.014$), but no effects of Task Load. For the low spatial frequency, there were significant effects of Deviance ($F(1,13)=9.83$, $p<0.008$), Task Load ($F(2,12)=20.30$, $p<0.002$), and their interaction ($F(2,12)=5.19$, $p<0.024$). The interaction indicated that the deviance effect was larger for low spatial frequencies in the easier task load conditions. Hence, the absence of interactions with spatial frequency in the three-way analysis was due to the fact that the modulation of the deviance effect by spatial frequency occurred only in the "easy" task load condition to mitigate the statistical outcomes. As a check on chance capitalization, additional analyses were performed on the Oz data in the bottom panel of Figure 6 that confirmed the earlier results (i.e., for both the high and low spatial frequency, the main effect of Deviance was significant, but the Task Load and Deviance × Task Load effects were not). Thus, the differential spatial frequency effects for the Fz data found in the separate analyses appear to be genuine effects and not Type I errors, and the absence of the interaction in the three-way analysis due to low statistical power (a Type II error).

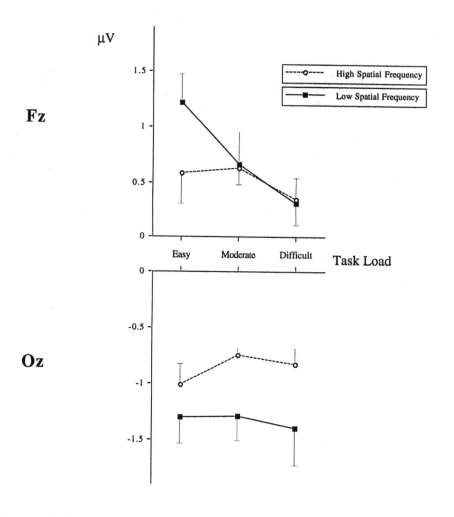

Figure 6. Deviance effects at Fz (upper panel) and Oz (lower panel) between 120 and 180 milliseconds. Shown are the means and standard errors of the difference between responses to standard and deviant spatial frequencies, separately for each spatial frequency and task load condition (n=14).

4. DISCUSSION

Task-irrelevant gratings with either low or high spatial frequencies were flashed to the upper and lower part of a computer screen while subjects performed a continuous compensatory tracking task at three levels of difficulty. In half of the blocks, the low spatial frequency grating was standard and the high spatial frequency grating was deviant, in the other half this mapping was reversed. This design isolated effects related to the physical difference between the stimuli, effects related to the deviance of the stimuli, and effects related to interactions and modulations. Such a design is necessary if effects of infrequent presentation probability are to be interpreted as true deviance effects, without the confounding effects of different exogenous stimulus parameters. The logic and the design requirements here are identical to those in studies of voluntary attention, with stimulus relevance replaced by stimulus deviance (Heslenfeld et al., 1997).

The performance data showed that the three difficulty levels of the compensatory tracking task had the intended effects: keeping the moving rectangle in the middle of the screen was significantly easier in the "easy" condition and significantly more difficult in the "difficult" condition than it was in the "moderate" condition. In addition and at the more difficult levels only, frequent stimulation with low spatial frequencies was more interfering with the tracking task than frequent stimulation with high spatial frequencies. This result is highly consonant with evidence showing that lower spatial frequencies are more capable of interrupting and interfering with ongoing visual processing than higher spatial frequencies. This evidence stems from neurophysiological and psychophysical data showing that lower spatial frequencies are preferably transmitted through transient visual channels that are thought to have larger inhibitory effects than the more sustained visual channels, which preferably transmit higher spatial frequencies (cf. Blum, 1991; Breitmeyer & Ganz, 1976; Hughes, 1986; Hughes et al., 1996).

With respect to the evoked potentials, high spatial frequency gratings evoked a much larger early C1 component (60-100 milliseconds) than low spatial frequency gratings, which is in line with the well-known effects of spatial frequency on this component (Smith & Jeffreys, 1978; Spekreijse et al., 1973). This difference was unaffected by task load, which shows that visual task difficulty did not alter the early exogenous representations of these stimuli. This result is important because it excludes a number of trivial factors (such as differences in fixation, accommodation, arousal, etc.) as possible causes of other effects involving task load.

However, a significant effect of stimulus deviance on the C1 component to high spatial frequency gratings was observed (see Figure 5). Deviant high spatial frequencies evoked a larger C1 response than standard high spatial frequencies. The latency, scalp distribution, and spatial frequency specificity of this effect clearly links it to the exogenous C1. It can be interpreted as a saturation of the response to high spatial frequency gratings when these stimuli were frequently presented, as opposed to a full response when these stimuli were deviant. This result is again very much in line with the different temporal characteristics of transient and sustained visual channels. Sustained channels, transmitting higher spatial frequencies, are thought to have slower conduction velocities and longer recovery times than transient channels (Breitmeyer & Ganz, 1976; Stone, 1983; cf. Brannan, 1992). Thus, responses to high spatial frequency gratings may saturate earlier than responses to low spatial frequency gratings when the respective stimuli are more frequently presented.

This saturation effect was unaffected by task load, which again demonstrates the immunity of these early exogenous responses to the difficulty of the overt visual tracking task, as well as their replicability across task blocks. This finding further strengthens the interpretation of these effects as reflecting purely exogenous processes and again excludes trivial differences in fixation or arousal as possible causes of other task load effects.

At frontal and central scalp sites, a positive response to deviant stimuli (120-180 milliseconds) was found, which depended on both task load and spatial frequency. That is, the response was larger to low spatial frequency deviants in the "easy" task load condition. The spatial frequency dependence of this response is again in line with the different functional characteristics of transient and sustained visual channels. Transient channels, transmitting lower spatial frequencies, are ascribed larger interruptive and attention-capturing powers than sustained channels (Brannan, 1992; Breitmeyer & Ganz, 1976; Hughes et al., 1996), which may have led to an enhanced or further processing of the low spatial frequency deviants. This effect may also be related to the frontal component of the auditory MMN (Giard et al., 1990), which is thought to reflect an involuntary shift of attention towards unexpected, deviant input (Näätänen, 1992; see also Alho et al., 1998). As the present data suggest, this shift may be stronger (or may occur more frequently) the more processing resources available, and the more salient the unexpected deviant input.

At posterior scalp sites, a negative response to deviant stimuli (120-200 milliseconds) was obtained, which was dissociated from the frontal effect by its independence of both spatial frequency and task load. This response was initially largest at the occipital midline (Oz, 120 to about 160 milliseconds), and later largest at lateral occipito-temporal sites (T5/T6, from about 160 to 200 milliseconds; see Figure 4), which shows that it consisted of two separable sets of generators. However, both were independent of task load and spatial frequency, such that both were independent of the concurrent visual tracking task difficulty and of the evoking deviant stimulus physical characteristics. Both were also independent of the possibly reduced quality of the exogenous stimulus representation as indicated by the attenuation of the earlier C1 response specifically to high spatial frequency gratings. Thus, both responses may be interpreted as reflections of endogenous (feature-independent) and automatic (attention-independent) processing of deviant visual input. As such, they may be interpreted as the visual analogues of the auditory mismatch negativity (MMN).

The present visual MMN differs from the auditory MMN in that the former seems to be completely independent of the features of the evoking stimulus. In the auditory modality, the orientations of the generators in the upper temporal cortex have been reported to depend on the pitch of the evoking deviant tone (Tiitinen et al., 1993). These findings imply that the auditory MMN may be less endogenous than the visual MMN. However, the feature dependence of the auditory MMN is not firmly established, because this effect (a) was apparently not replicated in a subsequent study (Tiitinen et al., 1994), and (b) may have been caused by a summation of highly feature-specific N1 generators and less feature-specific MMN generators (Alho, 1995). Future research in both modalities is needed to clarify the exact extent to which both MMNs are truly independent of the features of the evoking stimulus, as well as resistant to more severe withdrawals of modality-specific attentional resources.

REFERENCES

Alain, C., & Woods, D.L. (1997). Attention modulates auditory pattern memory as indexed by event-related potentials. *Psychophysiology, 34,* 534-546.

Alho, K. (1995). Cerebral generators of mismatch negativity (MMN) and its magnetic counterpart (MMNm) elicited by sound changes. *Ear and Hearing, 16,* 38-51.

Alho, K., Winkler, I., Escera, C., Huotilainen, M., Virtanen, J., Jääskeläinen, I.P., Pekkonen, E., & Ilmoniemi, R.J. (1998). Processing of novel sounds and frequency changes in the human auditory cortex: Magnetoencephalographic recordings. *Psychophysiology, 35,* 211-224.

Alho, K., Woods, D.L., Algazi, A., & Näätänen, R. (1992). Intermodal selective attention. II. Effects of attentional load on processing of auditory and visual stimuli in central space. *Electroencephalography and Clinical Neurophysiology, 82,* 356-368.

Blum, B. (Ed.). (1991). *Channels in the visual nervous system: Neurophysiology, psychophysics and models.* London: Freund Publishing House.

Brannan, J.R. (Ed.). (1992). *Applications of parallel processing in vision*. Amsterdam: North Holland.

Breitmeyer, B.G., & Ganz, L. (1976). Implications of sustained and transient channels for theories of visual pattern masking, saccadic suppression, and information processing. *Psychological Review, 83,* 1-36.

Czigler, I., & Csibra, G. (1990). Event-related potentials in a visual discrimination task: Negative waves related to detection and attention. *Psychophysiology, 27,* 669-676.

Czigler, I., & Csibra, G. (1992). Event-related potentials and the identification of deviant visual stimuli. *Psychophysiology, 29,* 471-485.

Giard, M.H., Perrin, F., Pernier, J., & Bouchet, P. (1990). Brain generators implicated in the processing of auditory stimulus deviance: A topographic event-related potential study. *Psychophysiology, 27,* 627-640.

Harter, M.R., Condor, E.S., & Towle, V.T. (1980). Orientation-specific and luminance effects: Interocular suppression of visual evoked potentials in man. *Psychophysiology, 17,* 141-145.

Harter, M.R., & Guido, W. (1980). Attention to pattern orientation: Negative cortical potentials, reaction time, and the selection process. *Electroencephalography and Clinical Neurophysiology, 49,* 461-475.

Harter, M.R., Towle, V.L., & Musso, M.F. (1976). Size specificity and interocular suppression: Monocular evoked potentials and reaction times. *Vision Research, 16,* 1111-1117.

Heslenfeld, D.J., Kenemans, J.L., Kok, A., & Molenaar, P.C.M. (1997). Feature processing and attention in the human visual system: an overview. *Biological Psychology, 45,* 183-215.

Hughes, H.C. (1986). Asymmetric interference between components of suprathreshold compound gratings. *Perception and Psychophysics, 40,* 241-250.

Hughes, H.C., Nozawa, G., & Kitterle, F. (1996). Global precedence, spatial frequency channels, and the statistics of natural images. *Journal of Cognitive Neuroscience, 8,* 197-230.

McCarthy, G., & Wood, C.C. (1985). Scalp distributions of event-related potentials: An ambiguity associated with analysis of variance models. *Electroencephalography and Clinical Neurophysiology, 62,* 203-208.

Näätänen, R. (1990). The role of attention in auditory information processing as revealed by event-related potentials and other brain measures of cognitive function. *Behavioral and Brain Sciences, 13,* 201-288.

Näätänen, R. (1992). *Attention and brain function*. Hillsdale, NJ: Lawrence Erlbaum Associates.

Näätänen, R. (1995). The mismatch negativity: A powerful tool for cognitive neuroscience. *Ear and Hearing, 16,* 6-18.

Nyman, G., Alho, K., Laurinen, P., Paavilainen, P., Radil, T., Reinikainen, K., Sams, M., & Näätänen, R. (1990). Mismatch negativity (MMN) for sequences of auditory and visual stimuli: Evidence for a mechanism specific to the auditory modality. *Electroencephalography and Clinical Neurophysiology, 77,* 436-444.

Porges, S.W., & Bohrer, R.E. (1990). Analyses of periodic processes in psychophysiological research. In J.T. Cacioppo & L.G. Tassinary (Eds.), *Principles of psychophysiology: Physical, social, and inferential elements* (pp. 708-753). New York: Cambridge Press.

Ritter, W., Deacon, D., Gomes, H., Javitt, D.C., & Vaughan, H.G. (1995). The mismatch negativity of event-related potentials as a probe of transient auditory memory: A review. *Ear and Hearing, 16,* 52-67.

Sams, M., Paavilainen, P., Alho, K., & Näätänen, R. (1985). Auditory frequency discrimination and event-related potentials. *Electroencephalography and Clinical Neurophysiology, 62,* 437-448.

Smith, A.T., & Jeffreys, D.A. (1978). Size and orientation specificity of transient visual evoked potentials in man. *Vision Research, 18,* 651-655.

Spekreijse, H., van der Tweel, L.H., & Zuidema, T. (1973). Contrast evoked responses in man. *Vision Research, 13,* 1577-1601.

Stone, J. (1983). *Parallel processing in the visual system.* New York: Plenum Press.

Tales, A., Newton, P., Troscianko, T., & Butler, S. (1999). Mismatch negativity in the visual modality. *Neuroreport, 10,* 3363-3367.

Tiitinen, H., Alho, K., Huotilainen, M., Ilmoniemi, R.J., Simola, J., & Näätänen, R. (1993). Tonotopic auditory cortex and the magnetoencephalographic (MEG) equivalent of the mismatch negativity. *Psychophysiology, 30,* 537-540.

Tiitinen, H., May, P., Reinikainen, K., & Näätänen, R. (1994). Attentive novelty detection in humans is governed by pre-attentive sensory memory. *Nature, 372,* 90-92.

Wickens, C.D. (1984). Processing resources in attention. In R. Parasuraman & R. Davies (Eds.), *Varieties of attention* (pp. 63-102). New York: Academic Press.

Woestenburg, J.C., Verbaten, M.N., & Slangen, J.L. (1983). The removal of eye-movement artefact from the EEG by regression analysis in the frequency domain. *Biological Psychology, 16,* 127-147.

Woldorff, M.G., Hackley, S.A., & Hillyard, S.A. (1991). The effects of channel-selective attention on the mismatch negativity wave elicited by deviant tones. *Psychophysiology, 28,* 30-42.

Woldorff, M.G., Hillyard, S.A., Gallen, C.C., Hampson, S.R., & Bloom, F.E. (1998). Magnetoencephalographic recordings demonstrate attentional modulation of mismatch-related neural activity in human auditory cortex. *Psychophysiology, 35,* 283-292.

Woods, D.L., Alho, K., & Algazi, A. (1992). Intermodal selective attention. I. Effects on event-related potentials to lateralized auditory and visual stimuli. *Electroencephalography and Clinical Neurophysiology, 82,* 341-355.

Chapter 4

CHANGE DETECTION IN COMPLEX AUDITORY ENVIRONMENT: BEYOND THE ODDBALL PARADIGM

ISTVÁN WINKLER
Institute of Psychology, Hungarian Academy of Sciences, Cognitive Brain Research Unit, Department of Psychology, Helsinki University

New auditory information usually appears as a change in some parameter of the acoustic input. Detecting and processing change are, therefore, important functions of the human auditory system. Studying how the human brain processes acoustic change has a long tradition in psychology (e.g., James, 1890). However, in complex natural auditory environments, it is not easy to determine what constitutes change. In the present review, "change" will be regarded as deviation from an aspect of the acoustic environment that *has been registered by the system* as regular input. This definition of change includes the notion that for any given system, including living organisms, change always corresponds to some previously registered regularity. Violation of a regularity that was not identified as such by a given system does not constitute change (for this system). For example, the human auditory system cannot usually detect long, periodically repeating sound patterns due to capacity limitations of the auditory sensory memory store (Guttman & Julesz, 1963). As a consequence, one cannot detect violations of such regularities. Another problem for conceptualizing change in complex auditory scenes is that a sound may simultaneously violate several regularities. To determine which aspect (or aspects) of this sound activated change-related processes, one must study what kind of auditory regularities were represented in the brain. Often, when the experimenter presents an auditory stimulus sequence to subjects and measures responses elicited by acoustic change, the contents of the regularity underlying the change-related response seem to be obvious to the experimenter since these experimental situations are usually very simple. However, the human brain is geared to deal with much more complex auditory environments and, therefore, might treat the sequence somewhat differently from what the experimenter expects.

The goal of this review is to show that the processes involved in detecting changes in natural auditory environments can be assessed by using relatively simple paradigms. The evidence suggests that even in such simplified acoustic situations the human auditory system goes beyond what seems "obvious" to the experimenter in terms of identifying and representing regularities. The discussion focuses on capabilities of stimulus-driven (bottom-up) processing of auditory change and is based primarily on neuroimaging methods and especially event-related brain potentials (ERP), which have provided important insights into stimulus-driven processing of auditory information.

1.   ERP MEASURES OF STIMULUS-DRIVEN CHANGE DETECTION

As an initial paradigmatic example of methods to study change detection, subjects are presented with a sound sequence consisting of a single repeating tone (termed the standard). Occasionally, the standard is exchanged for a different tone (termed the deviant), for example, one having a higher frequency—a task procedure often called the oddball paradigm. It has been known for more than 20 years that the deviant stimulus elicits an ERP component, termed the mismatch negativity (MMN, Näätänen, Gaillard, & Mäntysalo, 1978; for a review, see Näätänen & Winkler, 1999) whether or not the subject's task requires attention to be focused on the auditory stimuli. Highly distinguishable (salient) deviant sounds also elicit another ERP component termed P3a (Squires et al., 1975; for reviews, see Knight & Scabini, 1998; Polich, 1998).[1] As neither MMN nor P3a are elicited by regular or repetitive sounds, these components appear to index the neural activity involved in processing stimulus change.

It is important to note that MMN and P3a reflect different processes and can be dissociated from each other. MMN is not always followed by a P3a, but MMN is always followed by a P3a when the deviant stimulus is salient in the given auditory environment (Lyytinen et al., 1992). Moreover, even when both MMN and P3a are elicited in an auditory oddball paradigm, the amplitudes of the two components do not always vary together (Winkler et al., 1998). P3a can also be elicited without a corresponding MMN. An example that would elicit this component would be a loud or otherwise salient sound presented after a long silent period (Woods, 1990). However, only sounds deviating from some previously established non-silent auditory regularity elicit the MMN. Based on results from a large number of studies, current theories suggest that, whereas MMN is likely involved in processing auditory change within the large-capacity, stimulus-driven processing system (Näätänen, 1990, 1992; Näätänen & Winkler, 1999; Winkler et al., 1996), P3a might reflect the redirection of attention to a stimulus that was encountered outside the focus of attention or "passive attention" (James, 1890; see also, Näätänen, 1990, 1992; Öhman, 1979).

The oddball paradigm is a model of changes that occur in natural acoustic situations. However, even though quite simple in structure, one might question whether it is necessary to assume the existence of a system pre-attentively representing auditory regularities as the basis of MMN elicitation. An alternative account suggests that the frequent repetition of the standard stimulus creates an unnaturally sharp sensory memory trace of this sound in the auditory system. MMN is elicited either by an automatic process comparing the trace of the deviant sound with the standard-stimulus memory trace (Ritter et al., 1995), or by the process that forms the sensory memory trace of the deviant sound in the presence of a strong standard-stimulus trace (Näätänen, 1984).

Studies showing that MMN can be elicited without repeating any given sound, argue against this "strong memory trace" explanation of MMN. Sound repetition is not a necessary prerequisite of the change detection process indexed by the MMN. Tervaniemi, Maury, and Näätänen (1994) presented tone sequences that were either ascending or descending in frequency. Occasional tones the frequency of which did not fit into the trend elicited the MMN. This result showed that the regularity extracted from sequences was based on the direction of frequency change between successive tones, not on the repetition of some sound (Tervaniemi et al., 1994). Other paradigms established that abstract rules can also provide a basis for the change detection reflected by the MMN (Paavilainen et al., 1995, 1999; Saarinen et al., 1992). For example, when "the higher the frequency the lower the intensity" rule applied to the majority of the tones in a sequence, occasional deviants with low frequency and low intensity (or high frequency and high intensity) elicited the MMN (Paavilainen et al., 2001). This result again cannot be explained solely on the basis of auditory sensory memory traces of some sounds.

Furthermore, MMN may not be elicited by deviants following a long (12 seconds) silent interval, even when subjects are able to discriminate the standard and deviant tones across this time period (Berti et al., 2000). This observation can be linked with those showing that MMN is not elicited by a change between two sounds without the first sound being repeated a few times, although the sensory memory representations are sufficiently clear for voluntary discrimination of the sounds (Cowan et al., 1993). Instead, the same sound must be presented a few (minimally 3) times for a subsequent different sound to elicit the MMN (Cowan et al., 1993; Horváth et al., 2001; Schröger, 1997; Winkler et al., 1996a and b). That is, the sensory representation of an auditory stimulus can only serve as the "standard" for the MMN-generating change detection process if this stimulus is repeated prior to (but not long before) the "deviant" sound.

The results reviewed above strongly support the notion that the change-related processes reflected by the MMN component are based on representations of the current auditory regularities, rather than on strong sensory memory traces of a repeating sound (Winkler et al., 1996b). Even in the case of a single repeating tone, as in the auditory oddball paradigm, the "standard" underlying change detection is a regularity (the repetition of a tone, including the auditory representation of the tone itself), not just the auditory sensory memory trace of the repeating tone.

Because the auditory oddball paradigm and the most commonly used methods of sound presentation neglect so many aspects of natural situations, the question arises as to whether MMN can be elicited in complex natural acoustic environments or whether it is the product of an artificially constrained stimulation procedure used in the experiments in which the MMN component was originally discovered. In other words, does MMN reflect an important part of the change-detection system processing new information in everyday life, or is it more or less a laboratory artifact?[2] The following sections will discuss the features of natural auditory scenes that are missing from the typical oddball paradigm, and will review results of experiments that were designed to model these natural features.

2.   STIMULUS PRESENTATION

There are two aspects of auditory stimulus presentation used in most ERP experiments, which neglect important characteristics of natural acoustic environments: sound complexity and free-field sound delivery. Most studies present simple sinusoid tone bursts, whereas sounds encountered in natural situations are usually more complex. Addressing this issue, studies of harmonic tones (Tervaniemi et al., 2000; Winkler et al., 1997), chords (Alho et al., 1996), synthesized and digitized natural speech sounds (e.g., Sams et al., 1990; Sandridge, & Boothroyd, 1996), noise bursts (Nordby et al., 1994), and other complex sounds (Sams & Näätänen, 1991; Winkler et al., 1998) have shown that MMN as well as the P3a are elicited (in fact they might even be larger in amplitude, see, e.g., Csépe & Molnár, 1997; Tervaniemi et al., 2000) when complex sounds are used. One should note that natural MMN-eliciting sounds often also elicit P3a and natural (environmental or complex harmonic) sounds embedded in sequences of simple tone bursts invariably evoke this component (Escera et al., 2000).

General ERP research practice includes the use of headphones to deliver sound, which permits fine control over auditory parameters. However, this technique is not a necessary prerequisite of MMN or P3a elicitation, as a number of studies have demonstrated the elicitation of change-related ERP components using stimuli presented via loudspeakers (e.g., Paavilainen et al., 1989; Winkler et al., 1998). Thus, the auditory change ERP components are elicited when sounds are presented in a natural way.

3.   ACOUSTIC VARIANCE

The most obvious abstraction of the oddball paradigm (compared to natural situations), is that the standard stimuli are identical sounds presented isochronously. Very few natural sound sources occur this way. Even with such a sound source, one would still have to assume a total lack of variance in the position of the listener as well as all other objects of the given environment that could affect acoustic reflections. Therefore, if the change-related processes reflected by MMN and P3a are natural phenomena, they should tolerate acoustic variance.

Figure 1 provides illustrative examples of this assertion. Winkler et al., (1990) found that MMN is elicited when one tonal feature (intensity) of the standard stimulus was varied. Deviants differing from the standard in either the varying feature intensity (Figure 1a) or another feature (frequency, Figure 1b) elicited MMN. The MMN's tolerance to acoustic variance has been demonstrated with infrequent changes in a constant sound feature when several other features vary (Houtilainen et al., 1993; Gomes et al., 1995). As outlined above, several studies demonstrated that abstract regularities, such as frequency ascension (or descension), can be pre-attentively extracted from the auditory input (Paavilainen et al., 1995, 1999, 2001; Saarinen et al., 1992). This means that the auditory regularity representation system underlying MMN generation tolerates acoustic variance and can use it to extract higher-order regularities. These findings suggest a primitive sensory intelligence that is active even when the sounds lie outside the focus of attention.

Another source of acoustic variance stems from the timing of sound delivery. Although, most experimental designs used uniform sound presentation rates, constancy of stimulus delivery rate is not necessary for

MMN elicitation in the oddball paradigm (e.g., Böttcher-Gandor & Ullsperger, 1992). In addition, Horváth et al. (2001) showed that infrequently violating the alternation of two tones differing in frequency resulted in MMN elicitation even when the inter-stimulus interval between successive tones was varied. In sum, the change related processes reflected by the MMN response tolerate, or can even utilize variability in spectral as well as temporal acoustic parameters.

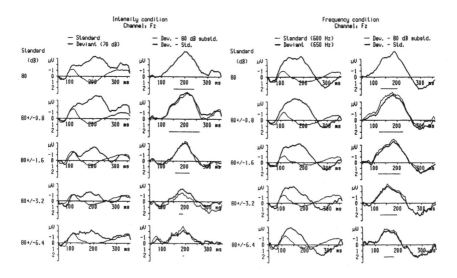

Figure 1. Grand-average (10 subjects) frontal (Fz) MMN responses elicited by intensity-deviant (70 dB/600 Hz, Figure 1a) and frequency-deviant (80 dB/650 Hz, Figure 1b) tones (p=0.10) presented amongst varying-intensity 600 Hz standard tones (p=0.90). The five levels of intensity variation (no, ±0.8, ±1.6, ±3.2, and ±6.4 dB) were tested in separate blocks using 9 equidistant intensity steps in each stimulus block (except in the no variance block). The 9 intensity variants of the standard tone ("substandards") were delivered with equal (p=0.10) probability. Responses to the standard (averaged across all different intensity levels; thin line) and deviant tones (thick line) are presented on the left side of the figure. Differences between the responses to the deviant and standard tones (averaged across all intensity variants; thick line) and those between the deviant and the 80 dB substandard tones (thin line) are shown on the right side of the figure. Different levels of intensity variation are shown as rows, separately for the intensity (A) and frequency (B) conditions (after Winkler et al., 1990).

4.   TEMPORAL PATTERNS VS. SINGLE SOUNDS

Natural sound sources usually emit patterns of sounds rather than single stimuli. This is another important feature in which the typical oddball paradigm simplifies natural situations. The acoustic regularities inherent in speech, music, and in most environmental sounds (e.g., the sound of the hooves of a galloping horse), can be better described in terms of temporal sound patterns, rather than in terms of individual sounds. The repetitive or otherwise regular feature of the sound sequence is based on segments consisting of a number of sound elements (e.g., the sequence of steps made up by the galloping horse), which may or may not be separated by silent intervals. For the regularity representation system to function efficiently in natural environments, the perceptual "units" and their regularities must be detected together as these often define the unit itself (Port, 1991).

Schröger and his colleagues (Schröger, 1994; Schröger, Näätänen, & Paavilainen, 1992; Schröger, Paavilainen, & Näätänen, 1994) demonstrated that MMN is elicited by occasionally changing the frequency of one tonal segment of a concatenated tonal pattern, consisting of 6-8 segments within a repetitive series of this tonal pattern, whether or not the tonal patterns were separated by silent intervals. Occasionally changing the frequency of a tonal segment introduces deviant frequency transitions at the borders of the deviant segment (even if the frequency of the deviant segment is equal to that of one of the other segments). Therefore, these results could also be explained by assuming that the auditory system encodes frequency changes between consecutive sounds (as was later confirmed by Paavilainen et al., 1999). To verify that MMN can be elicited by violating regularities based on tonal patterns rather than regularities based on individual tones or relationships between consecutive tones, Winkler and Schröger (1995) assessed whether MMN is elicited when two segments of identical frequency but different durations are exchanged. Furthermore, as illustrated in Figure 2a, the tonal patterns were presented under four different conditions: 1) as a sequence consisting of a discrete repeating tonal pattern, with consecutive patterns being separated by silent intervals, 2) as a periodic sequence repeating the same tonal pattern with no silent interval between consecutive patterns (two versions differing only in the duration of the tonal segments), and 3) as a sequence of four discrete tones repeating periodically with the same amount of silence separating consecutive tones and consecutive cycles of the four tones. Figure 2b indicates that MMN was obtained in all of these conditions, demonstrating that the regularity representations involved in the MMN-generating process can encode sound sequences in terms of patterns.

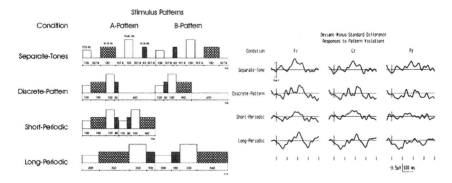

Figure 2. ERP responses to occasional violations of repeating tonal patterns. In each condition, two patterns of 4 tonal segments were used (Figure 2a). Pattern B was obtained by exchanging segments 2 and 4 of pattern A and vice verse. The exchanged tonal segments were identical in frequency but differed in duration. In half of the stimulus blocks of each condition, one pattern served as standard (p=0.90), the other as deviant (p=0.10). In the other half of the stimulus blocks the roles of the two patterns were exchanged. Timing and frequency parameters are shown on Figure 2a. Grand-average (12 subjects) frontal (Fz), central (Cz), and parietal (Pz) responses were calculated by subtracting the response to the standard pattern from the response to the same pattern when it served as the deviant (Figure 2b). The difference responses for the two patterns (A and B) were collapsed. The "reference" time point (marked on Figure 2b) was set to the onset of the difference between the standard and deviant patterns (i.e., the latency at which the shorter of the two exchanged segments ended). The tonal patterns (Figure 2a) and ERP difference responses (Figure 2b) for the 4 condition are shown in separate rows (after Winker et al., 1995).

Winkler and Schröger's (1995) results obtained in the separate-tone condition (item 3 above) bring up the question: When does the auditory system represent a sound sequence in terms of individual sounds and when does it do so in terms of sound patterns? What rules govern the formation of patterns? Scherg et al. (1989) reported that regular repetition of a pattern of tones is not sufficient in itself by presenting periodically repeated 5 tone sequences with 4 identical tones followed by a different one (SSSSD) and the inter-stimulus interval (ISI) being the same within and between patterns. MMN was elicited by the pattern-ending D tones just as it was when the S and D tones were presented in a randomized order with the same probabilities (0.80-0.20, respectively). These results suggested that in both conditions (regular and randomized-order presentation), the regularity encoded by the system underlying the MMN-generating process was the repetition of the S tone. D tones violated this regularity, thus eliciting MMN. If the regular sequence had been represented in terms of the repeating 5-tone pattern, the pattern-ending D tone should not have elicited MMN, as it would have been encoded as part of the regularity (i.e., a part of the repeating 5-tone standard pattern). Sussman et al. (1998a) hypothesized that the relatively slow presentation rate used in the regular-presentation condition (1.3 second stimulus onset asynchrony [SOA]) by Scherg et al. (1989) prevented pre-attentive detection of pattern repetition, since one full cycle pattern lasted 6.5 seconds. Hence, as three presentations of a pattern is the minimum required for establishing a regularity for MMN (Cowan et al., 1993; Schröger, 1997; Winkler et al., 1996b), these long auditory patterns exceeded by far the estimated temporal span of auditory sensory memory (ca. 10 seconds, Cowan, 1984). In contrast, in the separate-tone condition of Winkler and Schröger's (1995) study, one cycle of the tones lasted only 860 milliseconds. Sussman et al. (1998a) speeded up the rate of stimulus delivery in Scherg et al.'s paradigm to 100 milliseconds SOA (resulting in a 500 milliseconds cycle for the tone pattern) and found that the D tones no longer elicited MMN. However, the D tones did elicit MMN in the comparable randomized-order condition, so that the repetition of a tone pattern can only be detected without focused attention if the cycles are sufficiently short.[3] Thus, even though the capacity of the pre-attentive system for detecting temporal sound patterns might be limited compared to what can be attentively processed, stimulus-driven processing seems to be well prepared to deal with regularities based on sound patterns, as is required for detecting changes in natural situations.

5.   NON-REPETITIVE REGULARITIES

Regularities that may apply to a given source are not restricted to repetition of a sound or a sound pattern. This is another important feature of natural environments that is not modeled by the oddball paradigm. Smooth transitions in location or pitch are usually parts of the regularities, whereas abrupt changes may signal outstanding auditory events. A change detection system working under everyday conditions must be able to separate smooth transitions and encode them as regular features, so that the transitions are not confused with violations of non-repetitive regularities, like those found with feature trends (pitch: Tervaniemi et al., 1994; virtual movement in space: Winkler et al., in preparation), abstract relationships (Horváth et al., 2001; Paavilainen et al., 1995, 1999, 2001; Saarinen et al., 1992), and even repeating one sound in an ever-changing sequence of tones (Horváth et al., 2001; Wolff & Schröger, 2001) elicit MMNs. These results demonstrate that the processes underlying the MMN component can encode transition-based and higher-order (non-repetitive) regularities, separating them from irregular changes in sound features (which are detected as regularity violations).

6.   TEMPORAL DYNAMICS AND MULTIPLE REPRESENTATIONS OF REGULARITIES

In real life situations, the nature of what is regular might change in time. Some aspects of the stimulation that have been regular for some time may cease to be so or change their characteristics, while other aspects may become regular. Several changes may occur together, but they can also happen separately. Therefore, the human auditory change detection system must be able to dynamically adapt its regularity representations by eliminating outdated regularity representations and building up new ones.

Winkler et al. (1996b) studied the course of the MMN while the regularity of the sound sequence changed from one repetitive tone to another. Short trains of tones were presented to the subject. Each train started with 6 long tones (450 milliseconds in duration), establishing the long tone as a repetitive regularity. The long tones were followed by 2, 4, or 6 short tones (150 milliseconds in duration), or no short tones at all, as seen in Figure 3a. Each train ended with a medium-duration (300 milliseconds long) probe tone. Because the peak latency of the MMN response follows between 100-200 milliseconds from the time when deviation from the regularity commences, the latency of the MMN response elicited by the probe tone revealed which of the two possible standards (short or long) was active at the time the probe tone was presented. A MMN elicited by the long-tone standard could be expected to peak between 400 and 500 milliseconds from the onset of the probe tone, because the probe tone started to differ from the long tone at its offset (300 milliseconds). In contrast, the probe tone started to differ from the short tone at the offset of the short tone (150 milliseconds). Therefore, a MMN elicited by the probe with respect to the short tone could be expected to peak between 250 and 350 milliseconds from the onset of the probe. Figure 3b illustrates results from the probe elicited MMN with respect to the long-tone standard under three different circumstances: when no, 2 or 4 short standards intervened between the last long tone and the probe. The probe elicited MMN with respect to the short-tone standard with 4 or 6 short tones preceding it. When the 6 long tones were followed by 4 short ones, the probe elicited two MMNs, one with respect to each of the two standards. Naturally, the first few (2 or 3) short tones following the 6 long ones also elicit MMN with respect to the long-tone standard. These results demonstrated that the auditory system closely follows the change from one repeating tone to another.

Changing the Standard Stimulus

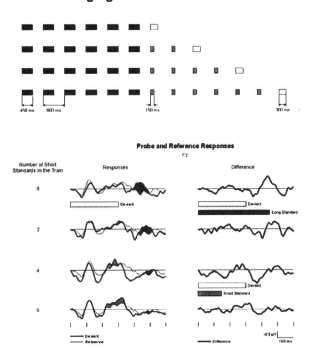

Figure 3. ERP responses elicited by medium-duration (300 milliseconds) probe tones following a change of the repeating (standard) stimulus. Four different short trains (Figure 3a) were equiprobably mixed together within the stimulus blocks. Trains were separated by the same interval (800 milliseconds onset-to-onset delay) as consecutive tones within the trains. Each train started with 6 long-duration (450 milliseconds) tones and ended by the probe tone. Trains differed in the number of short-duration (150 milliseconds) tones intervening between the last long-duration tone and the probe: either no, 2, 4, or 6 short tones were delivered (Figure 3a). The 3 types of tones differed only in their duration; other parameters (frequency 1000 Hz, intensity 80 dB) were identical for all tones. Figure 3b (left side) presents the grand-average (8 subjects) frontal (Fz) responses to probe (thick line) and identical reference tones (thin line). The reference response was obtained in a stimulus separate block in which the probe tone was presented alone with the stimulus delivery rate used in the other blocks. The right side of Figure 3b shows the probe-minus-reference difference responses. Since the MMN response usually follows by 100-200 milliseconds the moment when the difference between the deviant (probe) and the "standard" stimulus commences, the early frontally negative difference (peaking 250 and 350 milliseconds; marked by gray shading on the left side of Figure 3b) reflects an MMN elicited by the probe with respect to the short-duration standard stimulus. The late difference (peaking between 400 and 500 milliseconds; marked by black shading on the left side of Figure 3b) reflects an MMN elicited by the probe with respect to the long-duration standard stimulus. The long- and short-duration standards and the probe tone are marked on Figure 3b to help relating the MMN components to the deviance which elicited them. Responses elicited in different trains are presented in separate rows (after Winker et al., 1996).

Horváth et al. (2001) extended the notion of the adaptation of regularity representations to changing between two different types of regularities. These authors tested the transition from tone alternation (ABABAB...) to tone repetition (...BBB) and back to alternation. MMN was elicited by returning to alternation after repeating a tone twice (...ABABABBBA). This result suggests that the repetition rule for the B stimulus was established by just 3 consecutive presentations of this tone. The repetition rule was violated by the return of alternation (tone A). Repetitions of the B sound (...ABABABB and ...ABABABBB) also elicit MMN in this situation as they violate the preceding alternation rule. Regularly alternating tones (not immediately preceded by stimulus repetition) do not, of course, elicit MMN. Again, the new regularity (repeating the B tone) was represented soon after it appeared.

Taken together, these findings suggest that the change detection system reflected by MMN quickly adapts to the emergence of new regularities. The emergence of a new regularity does not in itself erase previous regularities. This was shown by the elicitation of two MMNs, one with respect to the long and the other with respect to the short-tone standard, by Winkler et al.'s (1996b) medium-duration probe tones. The elimination of an "outdated" regularity representation is somewhat slower. Several studies have found that regularity representations may be retained for quite long periods (>10 seconds), although they need to be reactivated before MMN can again be elicited with respect to them (Cowan et al., 1993; Ritter et al., 1998; Winkler et al., 1996a). Fast formation and relatively long retention of regularity representations are probably optimal strategies in natural auditory environments.

These characteristics of the pre-attentive regularity representation system demand simultaneous maintenance of multiple regularities for the same sound sequence. In the studies reviewed above, (Horváth et al., 2001; Winkler et al., 1996b) one regularity was exchanged for another. However, in many situations, the same sound sequence may have several different regular aspects. Even simple tone repetition might be regarded as a set of separate repetitive regularities for the different tone features, since infrequently violating the constancy of a given sound feature elicits MMN when other sound features are randomly varied in the sequence (Gomes et al., 1995; Huotilainen et al., 1993; Nousak et al., 1996; Winkler et al., 1990, 1995). The additivity between MMNs elicited by simultaneous deviance in some (but not all) combinations of tonal features (Levänen et al., 1993; Schröger, 1995; Takegata et al., 1999, 2001) partly support the view that feature repetitions are maintained in parallel to the repetition of the whole sound (Ritter et al., 1995).

Winkler and Czigler (1998) demonstrated simultaneous maintenance of two regularities for the same sound sequence, by occasionally delivering to subjects a deviant sound that violated two different auditory regularities of the test sequence. A sequence of two alternating tones was presented that differed only in frequency. Occasional deviants repeated the frequency of the previous tone (thus violating alternation) and were also shorter in duration than the standard (alternating) tones. The two successive MMN components elicited by these deviants indicated that both tone alternation and the constancy of tone duration were represented as regularities at the time when the deviant stimulus was delivered (for corroborating evidence, see Sussman et al., 1999). Moreover, even a simple rule, such as the alternation of two tones (ABABAB...) is simultaneously represented by several (redundant) regularities (Horváth et al., 2001). One rule enabled the auditory system to extrapolate from one tone to the next ("local" rule: "A" is followed by "B" and "B" is followed by "A"), the other from one tone to the tone following the next ("global" rule: every second tone is identical).[4] These results reveal that the machinery behind auditory change detection is a complex system forming, maintaining, and eliminating representations for multiple regularities even in seemingly simple auditory scenes (Bregman, 1990).

7.   MULTIPLE SIMULTANEOUSLY ACTIVE SOURCES

Finally, the auditory oddball paradigm reduces the variance present in natural environments by eliminating other sound sources. In real life, one seldom encounters situations in which only one sound source is active at a time. If human auditory change detection were unable to simultaneously handle regularities and deviations of multiple sound sources, it would either flood us with signals marking illusory changes (e.g., sounds that are regular in their own stream but violate some regularity of another stream) or only inform us about gross changes in the acoustic input.

Sussman et al. (1999) demonstrated that MMN can be made dependent on the segregation of two interleaved auditory sequences. These authors tested whether the processes of auditory streaming precede MMN generation. Auditory streaming is an important form of sound organization separating the signals of different sources, which appear together in the composite auditory input (Bregman, 1990; van Noorden, 1975). Streaming occurs when a sequence of sounds, mixed from two markedly different sound sets (e.g., high- and low-pitched tones), is presented at a fast pace. When it "streams" this sequence is perceived as two independent sound streams, and one can detect separate regularities in the two streams. When the physical separation between the two sets of sounds is small and/or the rate of stimulus delivery is slow, all sounds of the sequence are integrated into a single stream and regularities applying separately to the two sets of sounds cannot be perceived. Sussman et al. (1999a) showed that MMN elicitation follows the same pattern. High and low tones were alternated, repeating separate temporal tone patterns within the low and high sequence. When the sequence was presented at a slow pace, occasional changes in the high or low tone patterns did not elicit MMN. However, when the rate of tone delivery was increased, the same pattern violations elicited MMNs. Winkler et al., (submitted) showed that MMN elicitation and perception of the corresponding within-stream regularity appear together. More important, Ritter et al. (2000) have demonstrated that the regularities embedded in separate auditory streams are independent of each other. Therefore, deviant sounds elicit MMN only with respect to regularities of their own sound stream. These results indicate that the change detection processes indexed by MMN operate on an organized representation of the auditory input, in which multiple sound sources are processed in parallel and while maintaining separate independent regularity representations.

8.   FEATURES OF THE CHANGE DETECTION SYSTEM REFLECTED BY MMN

The change detection processes involved in MMN generation can operate in complex natural auditory environments. None of those characteristics of natural situations (which were not modeled by the auditory oddball paradigm) prevent the elicitation of MMN. MMN can be elicited by natural sounds presented in open acoustic fields. Variations in sound parameters are tolerated, even utilized by the regularity representation system underlying MMN generation. Temporal sound patterns can be registered as stimulus units in detected regularities. Non-repetitive regularities are identified and represented. Smooth transitions are identified as regularities, abrupt changes as regularity violations. The system follows and adapts to the temporal dynamics of regularities, usually tracking several regularities simultaneously. Signals from different sound sources are processed independently of each other. The fact that MMN indexes processes that have all these characteristics makes it a very useful tool for investigating auditory change detection.

9.   DISCUSSION

Although it is probably not true that all the processes up to and including MMN would be completely resistant to attentional manipulations, neither do they require one's attention to be focused on the sounds or the engagement of voluntary strategies for finding acoustic changes (Näätänen et al., 1993; Ritter et al., 1999; Sussman et al., 1998b; Woldorff et al., 1991). This feature of the processes indexed by MMN is very important in everyday life situations, since one part of the environment can be attended while effortlessly monitoring the rest of the auditory input. The system keeps the internal representation up-to-date and produces detection of sudden salient changes outside the current focus of attention. The processes creating an orderly representation of the auditory environment, segregating sources, grouping together sound segments, finding temporal and spectral regularities, even operate outside the focus of attention. However, these processes are not fully automatic as, at least in some cases, their outcome can be affected by top-down processes (Bregman, 1990; Sussman et al., 1998b). Such processes, which are neither strictly automatic nor fully dependent on attention, might be termed "default processes", analogous to those computer functions that do their job even without user interaction, but might be altered by entering explicit commands or parameters.

Change detection in natural auditory environments requires the existence of regularity representations. Forming and maintaining these representations are part of the processes that organize auditory input. The resulting set of analyzed information (containing descriptions of the sources and their current emission patterns) can be regarded as an internal model of the auditory environment. This model is essential for detecting changes in the unattended part of the auditory scene, as well as for selecting parts of the auditory scene. Due to the nature of the acoustic modality and the human auditory sensory organs, to select a single sound all other temporally overlapping sounds must also be delineated from the composite input. Otherwise, one would not be able to determine which part of the input belongs to the selected source. The fact that attention can be so quickly shifted between sources suggests that information about unattended sources is readily available.

The processes generating the MMN might play an important part in maintaining the internal model of the auditory environment. When a previously registered regularity is violated, two types of actions might be necessary: 1) calling for further processing of the new information and 2) updating the affected regularities. The first process may lead to a stimulus-initiated attention switch, shown by the elicitation of the P3a. The second process initiates changes in the internal model. Winkler et al. (1996b) suggested that MMN is generated by this latter process. Winkler and Czigler (1998) confirmed this hypothesis by showing that the number of MMNs elicited by a single deviant sound corresponds to the number of regularities that this sound violated within a short time period. That is, the same deviant may elicit one MMN by violating the same regularity twice or two MMNs if it violated two different regularities. Therefore, when one studies auditory change detection using the MMN component, one probes an essential part of the change detection process: the maintenance of acoustic regularities.

10.   SUMMARY

Processes of stimulus-driven auditory change detection can be studied using relatively simple stimulus paradigms. ERP components, such as the MMN and P3a, provide useful tools for investigating this important function of the human brain. However, the human auditory system is prepared for complex natural auditory environments and can do far more than what is required by any simplified test situation.

ACKNOWLEDGMENTS

The author is grateful to Dr. István Czigler, Dr. Erich Schröger, and Dr. Elyse Sussman for their comments on early versions of the manuscript. This work was supported by the Hungarian National Research Fund (OTKA 022681).

Figure 1 reprinted from Winkler, I., Paavilainen, P., Alho, K., Reinikainen, K., Sams, M., & Näätänen, R. (1990). The effect of small variation of the frequent auditory stimulus on the event-related brain potential to the infrequent stimulus. *Psychophysiology, 27,* 228-235. Copyright (2002), with permission from Cambridge University Press.

Figure 2 reprinted from Winkler, I., & Schröger, E. (1995). Neural representation for the temporal structure of sound patterns. *NeuroReport, 6,* 690-694. Copyright (2002) with permission from Lippincott Williams and Wilkins.

Figure 3 reprinted from Winkler, I., Karmos, G., & Näätänen, R. (1996b). Adaptive modeling of the unattended acoustic environment reflected in the mismatch negativity event-related potential. *Brain Research, 742,* 239-252. Copyright (2002) with permission from Elsevier Science.

NOTES

1 When the tones are attended, the deviant also elicits N2b (Renault & Lesévre, 1978; Ritter, Simson, & Vaughan, 1972; for a review, see Ritter & Ruchkin, 1992) and, if the subject's task is to find the deviant sounds, the P3 component (Sutton et al., 1965; Donchin & Coles, 1988).

2 Of course, there are good reasons to use simple sounds and paradigms in ERP research: to reduce the number of experimental variables and the noise of the measurements.

3 One could of course attentively detect the periodicity of Scherg et al.'s (1989) tone sequence.

4 Although maintaining multiple redundant representations may seem to be wasting resources, one should keep in mind that the alternation of two tones is a special case of a larger set of alternation regularities. There are alternation type regularities to which only one or the other rule may apply. For example, a sequence with a "higher pitch – lower pitch – higher pitch – lower pitch..." regular pattern is an alternating sequence, which conforms only to the above described "local" rule. Horváth et al. (2001) showed that occasional "higher-higher" and "lower-lower" segments embedded in such a sequence elicit MMN. Thus it seems that the pre-attentive auditory processing system is prepared to identify many different types of alternations, not only special cases (which are less likely to present themselves in natural situations).

REFERENCES

Alho, K., Tervaniemi, M., Huotilainen, M., Lavikainen, J., Tiitinen, H., Ilmoniemi, R.J., Knuutila, J., & Näätänen, R. (1996). Processing of complex sounds in the human auditory cortex as revealed by magnetic brain responses. *Psychophysiology, 33,* 369-375.

Berti, S., Schröger, E., Cowan, N., & Winkler, I. (2000). Attention and auditory sensory memory. In C. Escera, M. Tervaniemi, & E. Yago (Eds.), *Abstract book of the second International Congress on mismatch negativity and its clinical applications* (p. 67). Barcelona: University of Barcelona.

Böttcher-Gandor, C., & Ullsperger, P. (1992). Mismatch negativity in event-related potentials to auditory stimuli as a function of varying interstimulus interval. *Psychophysiology, 29,* 546-550.

Bregman, A.S. (1990). *Auditory scene analysis: The perceptual organization of sound.* Cambridge, MA: MIT Press.

Cowan, N., Winkler, I., Teder, W., & Näätänen, R. (1993). Memory prerequisites of the mismatch negativity in the auditory event-related potential (ERP). *Journal of Experimental Psychology: Learning, Memory, & Cognition, 19,* 909-921.

Csépe, V., & Molnár, M. (1997). Towards the possible clinical application of the mismatch negativity component of event-related potentials. *Audiology & Neuro-Otology, 2,* 354-369.

Donchin, E., & Coles, M.G.H. (1988). Is the P300 component a manifestation of context updating? *Behavioral and Brain Sciences, 11,* 357-374.

Escera, C., Alho, K., Schröger, E., & Winkler, I. (2000). Involuntary attention and distractibility as evaluated with event-related brain potentials. *Audiology & Neuro-Otology, 5,* 151-166.

Gomes, H., Ritter, W., & Vaughan, H.G., Jr. (1995). The nature of preattentive storage in the auditory system. *Journal of Cognitive Neuroscience, 7,* 81-94.

Guttman, N., & Julesz, B. (1963). Lower limits of auditory periodicity analysis. *Journal of the Acoustical Society of America, 35,* 610.

Horváth, J., Czigler, I., Sussman, E., & Winkler, I. (2001). Simultaneously active pre-attentive representations of local and global rules for sound sequences in the human brain. *Cognitive Brain Research, 12,* 131-144.

Huotilainen, M., Ilmoniemi, R.J., Lavikainen, J., Tiitinen, H., Alho, K., Sinkkonen, J., Knuutila, J., & Näätänen, R. (1993). Interaction between representations of different features of auditory sensory memory. *NeuroReport, 4,* 1279-1281.

James, W. (1890). *The principles of psychology.* New York: Holt.

Knight R.T, & Scabini D. (1998). Anatomic bases of event-related potentials and their relationship to novelty detection in humans. *Journal of Clinical Neurophysiology, 15,* 3-13.

Levänen, S., Hari, R., McEvoy, L., & Sams, M. (1993). Responses of the human auditory cortex to changes in one vs. two stimulus features. *Experimental Brain Research, 97,* 177-183.

Lyytinen, H., Blomberg, A.P., & Näätänen, R. (1992). Event-related potentials and autonomic responses to a change in unattended auditory stimuli. *Psychophysiology, 29,* 523-534.

Näätänen, R. (1984). In search of a short-duration memory trace of a stimulus in the human brain. In L. Pulkkinen & P. Lyytinen (Eds.), *Human action and personality. Essays in honor of Martti Takala. Jyväskylä Studies in education, psychology and social research* (pp. 29-43). Jyväskylä: University of Jyväskylä.

Näätänen, R. (1990). The role of attention in auditory information processing as revealed by event-related potentials and other brain measures of cognitive function. *Behavioral and Brain Sciences, 13,* 201-288.

Näätänen, R. (1992). *Attention and brain function.* Hillsdale, NJ: Lawrence Erlbaum Associates.

Näätänen, R., Gaillard, A.W.K., & Mäntysalo, S. (1978). Early selective attention effect on evoked potential reinterpreted. *Acta Psychologica, 42,* 313-329.

Näätänen, R., Paavilainen, P., Tiitinen, H., Jiang, D., & Alho, K. (1993). Attention and mismatch negativity. *Psychophysiology, 30,* 436-450.

Näätänen, R., & Winkler, I. (1999). The concept of auditory stimulus representation in cognitive neuroscience. *Psychological Bulletin, 125,* 826-859.

van Noorden, L.P.A.S. (1975). *Temporal coherence in the perception of tone sequences.* Unpublished doctoral dissertation, Eindhoven: University of Technology.

Nordby, H., Hammerborg, D., Roth, W.T., & Hugdahl, K. (1994). ERPs for infrequent omissions and inclusions of stimulus elements. *Psychophysiology, 31,* 544-552.

Nousak, J.M., Deacon, D., Ritter, W., & Vaughan, H.G., Jr. (1996). Storage of information in transient auditory memory. *Cognitive Brain Research, 4,* 305-317.

Öhman, A. (1979). The orienting response, attention and learning: An information-processing perspective. In H.D. Kimmel, E.H. van Olst, & J.F. Orlebeke (Eds.), *The orienting reflex in humans* (pp. 443-471). Hillsdale, NJ: Erlbaum.

Paavilainen, P., Jaramillo, M., Näätänen, R., & Winkler, I. (1999). Neural populations in the human brain extracting invariant relationships from acoustic variance. *Neuroscience Letters, 265,* 179-182.

Paavilainen, P., Karlsson, M.-L., Reinikainen, K., & Näätänen, R. (1989). Mismatch negativity to change in spatial location of an auditory stimulus. *Electroencephalography and Clinical Neurophysiology, 73,* 129-141.

Paavilainen, P., Saarinen, J., Tervaniemi, M., & Näätänen, R. (1995). Mismatch negativity to changes in abstract sound features during dichotic listening. *Journal of Psychophysiology, 9,* 243-249.

Paavilainen, P., Simola, J., Jaramillo, M., Näätänen, R., & Winkler, I. (2001). Preattentive extraction of abstract feature conjunctions from auditory stimulation as reflected by the mismatch negativity (MMN). *Psychophysiology, 38,* 359-365.

Polich J. (1998). P300 clinical utility and control of variability. *Journal of Clinical Neurophysiology, 15,* 14-33.

Port, R.F. (1991). Can complex temporal patterns be automatized? *Behavioral and Brain Sciences, 14,* 762-764.

Renault, B., & Lesévre, N. (1978). Topographical study of the emitted potential obtained after the omission of an expected visual stimulus. In D. Otto (Ed.), *Multidisciplinary perspectives in event-related brain potential research, EPA 600/9-77-043* (pp. 202-208). Washington: U.S. Government Printing Office.

Ritter, W., Deacon, D., Gomes, H., Javitt, D.C., & Vaughan, H.G., Jr. (1995). The mismatch negativity of event-related potentials as a probe of transient auditory memory: A review. *Ear and Hearing, 16,* 52-67.

Ritter, W., Gomes, H., Cowan, N., Sussman, E., & Vaughan, H.G., Jr. (1998). Reactivation of a dormant representation of an auditory stimulus feature. *Journal of Cognitive Neuroscience, 10,* 605-614.

Ritter, W., & Ruchkin, D.S. (1992). A review of event-related potential components discovered in the context of studying P3. In D. Friedman & G. Bruder (Eds.), *Psychophysiology and experimental psychopathology–A tribute to Samuel Sutton. Annals of the New York Academy of Sciences,* Vol. 658 (pp. 1-32). New York: The New York Academy of Sciences.

Ritter, W., Simson, R., & Vaughan, Jr., H.G. (1972). Association cortex potentials and reaction time in auditory discrimination. *Electroencephalography and Clinical Neurophysiology, 33,* 547-555.

Ritter, W., Sussman, E., Deacon, D., Cowan, N., & Vaughan, H.G., Jr. (1999). Two cognitive systems simultaneously prepared for opposite events. *Psychophysiology, 36,* 835-838.

Ritter, W., Sussman, E., & Molholm, S. (2000). Evidence that the mismatch negativity system works on the basis of objects. *Neuroreport, 11,* 61-63.

Saarinen, J., Paavilainen, P., Schröger, E., Tervaniemi, M., & Näätänen, R. (1992). Representation of abstract attributes of auditory stimuli in the human brain. *NeuroReport, 3,* 1149-1151.

Sams, M., Aulanko, R., Aaltonen, O., & Näätänen, R. (1990). Event-related potentials to infrequent changes in synthesized phonetic stimuli. *Journal of Cognitive Neuroscience, 2,* 344-357.

Sams, M., & Näätänen, R. (1991). Neuromagnetic responses of the human auditory cortex to short frequency glides. *Neuroscience Letters, 121,* 43-46.

Sandridge, S., & Boothroyd, A. (1996). Using naturally produced speech to elicit mismatch negativity. *Journal of the American Academy of Audiology, 7,* 105-112.

Scherg, M., Vajsar, J., & Picton, T.W. (1989). A source analysis of the late human auditory evoked potentials. *Journal of Cognitive Neuroscience, 1,* 336-355.

Schröger, E. (1994). An event-related potential study of sensory representations of unfamiliar tonal patterns. *Psychophysiology, 31,* 175-81.

Schröger, E. (1995). Processing of auditory deviants with changes in one versus two stimulus dimensions. *Psychophysiology, 32,* 55-65

Schröger, E. (1997). On the detection of auditory deviants: A pre-attentive activation model. *Psychophysiology, 34,* 245-257.

Schröger, E., Näätänen, R., & Paavilainen, P. (1992). Event-related potentials reveal how non-attended complex sound patterns are represented by the human brain. *Neuroscience Letters, 146,* 183-186.

Schröger, E., Paavilainen P., & Näätänen, R. (1994). Mismatch negativity to changes in a continuous tone with regularly varying frequencies. *Electroencephalography & Clinical Neurophysiology, 92,* 140-147.

Squires, K.C., Squires, N.K., & Hillyard, S.A. (1975). Decision-related cortical potentials during an auditory signal detection task with cued observation intervals. *Journal of Experimental Psychology: Human Perception and Performance, 1,* 268-279.

Sussman, E., Ritter, W., & Vaughan, H.G., Jr. (1998a). Predictability of stimulus deviance and the mismatch negativity. *NeuroReport, 9,* 4167-4170.

Sussman, E., Ritter, W., & Vaughan, H.G., Jr. (1998b). Attention affects the organization of auditory input associated with the mismatch negativity system. *Brain Research, 789,* 130-138.

Sussman, E., Ritter, W., & Vaughan, H.G., Jr. (1999a). An investigation of the auditory streaming effect using event-related brain potentials. *Psychophysiology, 36,* 22-34.

Sussman, E., Winkler, I., Ritter, W., Alho, K., & Näätänen, R. (1999b). Temporal integration of auditory stimulus deviance as reflected by the mismatch negativity. *Neuroscience Letters, 264,* 161-164.

Sutton, S., Braren, M., Zubin, J., & John, E.R. (1965). Evoked potential correlates of stimulus uncertainty. *Science, 150,* 1187-88.

Takegata, R., Paavilainen, P., Näätänen, R., & Winkler, I. (1999). Independent processing of changes in auditory single features and feature conjunctions in humans as indexed by the mismatch negativity. *Neuroscience Letters, 266,* 109-112.

Takegata, R., Paavilainen, P., Näätänen, R., & Winkler, I. (2001). Pre-attentive processing of simple and complex acoustic regularities: A mismatch negativity additivity study. *Psychophysiology, 38,* 92-98.

Tervaniemi, M., Ilvonen, T., Sinkkonen, J., Kujala, A., Alho, K., Huotilainen, M., & Näätänen, R. (2000). Harmonic partials facilitate pitch discrimination in humans: Electrophysiological and behavioral evidence. *Neuroscience Letters, 279,* 29-32.

Tervaniemi, M., Maury, S., & Näätänen, R. (1994). Neural representations of abstract stimulus features in the human brain as reflected by the mismatch negativity. *NeuroReport, 5,* 844-846.

Winkler, I., Cowan, N., Csépe, V., Czigler, I., & Näätänen, R. (1996a). Interactions between transient and long-term auditory memory as reflected by the mismatch negativity. *Journal of Cognitive Neuroscience, 8,* 403-415.

Winkler, I., & Czigler, I. (1998). Mismatch negativity: Deviance detection or the maintenance of the "standard". *NeuroReport, 9,* 3809-3813.

Winkler, I., Czigler, I., Jaramillo, M., Paavilainen, P., & Näätänen, R. (1998). Temporal constraints of auditory event synthesis: Evidence from ERPs. *NeuroReport, 9,* 495-499.

Winkler, I., Karmos, G., & Näätänen, R. (1996b). Adaptive modeling of the unattended acoustic environment reflected in the mismatch negativity event-related potential. *Brain Research, 742,* 239-252.

Winkler, I., Paavilainen, P., Alho, K., Reinikainen, K., Sams, M., & Näätänen, R. (1990). The effect of small variation of the frequent auditory stimulus on the event-related brain potential to the infrequent stimulus. *Psychophysiology, 27,* 228-235.

Winkler, I., & Schröger, E. (1995). Neural representation for the temporal structure of sound patterns. *NeuroReport, 6,* 690-694.

Winkler, I., Sussman, E., Tervaniemi, M., Ritter, W., Horváth J., & Näätänen, R. (submitted). Pre-attentive auditory context effects.

Winkler, I., Teder, W., Tervaniemi, M., & Näätänen, R. (in preparation). Brain electric responses to unexpected breaks in virtual sound movement.

Winkler, I., Tervaniemi, M., Huotilainen, M., Ilmoniemi, R., Ahonen, A., Salonen, O., Standertskjöld-Nordenstam, C-G., & Näätänen, R. (1995). From objective to subjective: Pitch representation in the human auditory cortex. *NeuroReport, 6,* 2317-2320.

Winkler, I., Tervaniemi, M., & Näätänen, R. (1997). Two separate codes for missing-fundamental pitch in the human auditory cortex. *Journal of the Acoustical Society of America, 102,* 1072-1082.

Winkler, I., Tervaniemi, M., Schröger, E., Wolff, Ch., & Näätänen, R. (1998). Pre-attentive processing of auditory spatial information: electrophysiological evidence from human subjects. *Neuroscience Letters, 242,* 49-52.

Woldorff, M.G., Hackley, S.A., & Hillyard, S.A. (1991). The effects of channel-selective attention on the mismatch negativity wave elicited by deviant tones. *Psychophysiology, 28,* 30-42.

Wolff, Ch., & Schröger, E. (2001). Activation of the auditory pre-attentive change detection system by tone repetitions with fast stimulation rate. *Cognitive Brain Research, 10,* 323-327.

Woods D.L. (1990). The physiological basis of selective attention: Implications of event-related potential studies. In J.W. Rohrbaugh, R. Parasuranam, & R. Johnson, Jr. (Eds.), *Event-related potentials: Basic issues and applications* (pp. 178-209). New York: Oxford University Press.

Chapter 5

THEORETICAL OVERVIEW OF P3a AND P3b

JOHN POLICH
Cognitive Electrophysiology Laboratory, Department of Neuropharmacology, The Scripps Research Institute, La Jolla, CA, USA

1.   P3a AND P3b

Recent empirical advances on the relationship between the P3a and P3b event-related brain potentials (ERPs) have suggested a plausible approach to how these potentials may interact. The purpose of this chapter is to review the issues surrounding these developments. The chapter is organized into several sections: First, the empirical background of the P3a and P3b subcomponent distinction is limned. Second, a theoretical perspective of P300 is presented. Third, the neuropsychological basis for the P300 component is outlined in terms of how these subcomponents may be related. The goal is to provide a theoretical overview of the topic areas by integrating prior findings with current perspectives.

1.1   Background

The P300 was discovered over 35 years ago and has provided much fundamental information on normal and dysfunctional cognition (Bashore & van der Molen, 1991; Sutton et al., 1965). Figure 1 illustrates how this ERP component is often elicited by using the "oddball" paradigm in the upper panel.   Two different stimuli are presented in a random order, and the subject is required to discriminate an infrequent target stimulus from the frequent standard stimulus by responding covertly or overtly to the target—typically a relatively easy discrimination (Picton, 1992; Polich, 1999). The target stimulus elicits the P300, which is not apparent in the ERP from the standard stimulus.

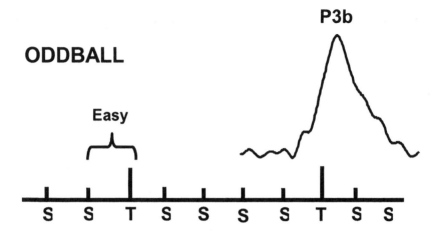

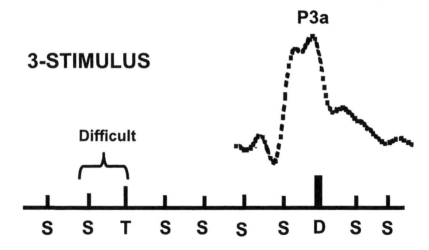

Figure 1. Schematic illustration of the oddball (upper panel) and three-stimulus (lower panel) paradigms, with the ERPs from the stimuli of each task presented at the right. The oddball task presents two different stimuli in a random sequence, with one occurring less frequently than the other (target =T, nontarget =N). The three-stimulus task also presents a compelling (not necessarily novel) distractor (D) stimulus that occurs infrequently, to which the subject does not respond but which elicits the "P3a" subcomponent. In each task, the subject responds only to the target stimulus, which elicits the "P3b".

The three-stimulus paradigm is a modification of the oddball task in which "distractor" stimuli are inserted into the sequence of target and standard stimuli. Figure 1 schematically illustrates the task situation in the lower panel. When "novel" stimuli (e.g., dog barks, color forms, etc.) are presented as distractors in the series of more "typical" target and standard stimuli (e.g., tones, letters of the alphabet, etc.), a P300 component that is large over the frontal/central areas can be produced with auditory, visual, and somatosensory stimuli (Courchesne et al., 1984; Knight, 1984; Yamaguchi & Knight, 1991b). This "novelty" P300 is sometimes called the "P3a" (Courchesne et al., 1975; Squires et al., 1975), and recent analyses confirm that these two potentials are the same brain potential (cf. Simons et al., 2001; Spencer et al., 1999). The parietal maximum P300 from the target stimulus is sometimes called the "P3b". As the P3a exhibits a frontal/central scalp distribution, relatively short peak latency, and rapidly habituates, it is thought to reflect frontal lobe function (Friedman & Simpson, 1994; Knight, 1997) and can be elicited in a variety of populations (Fabiani et al., 1998; Friedman et al., 1998; Yamaguchi & Knight, 1991a).

Infrequently presented nontarget visual stimuli that are easily recognized "typical" (i.e., not novel) also have been found to elicit a P300 with maximum amplitude over the central/parietal rather than frontal/central areas (Courchesne, 1978; Courchesne et al., 1978). This component is sometimes referred to as a "no-go" P300, because subjects do not respond to the infrequent nontargets (Falkenstein et al., 1995; Pfefferbaum et al., 1985). In the auditory modality, infrequent nontarget tone stimuli that are readily perceived (i.e., not novel) inserted into the traditional oddball sequence also elicit a central/parietal maximum P300 (cf. Katayama & Polich, 1996a; Pfefferbaum & Ford, 1988). When both an infrequent nontarget tone and a novel sound are presented, the novel stimuli elicit a central maximum P300 and the infrequent nontarget tone elicits a central/parietal P300, the amplitude of which is smaller than that of the novel stimulus potential (cf. Grillon et al., 1990; Verbaten et al., 1997). Thus, the P300 component can vary in amplitude and timing, because the intra-paradigm stimulus relationships define the stimulus context (cf. Katayama & Polich, 1996b; Suwazono et al., 2000).

1.2   Stimulus Context

Katayama and Polich (1998) assessed the role of task difficulty in the three-stimulus paradigm to examine stimulus context on the P300 scalp topography distribution. The perceptual distinctiveness between the target and standard stimuli was manipulated in an auditory task by using typical tone stimuli that varied in pitch. When the target/standard discrimination was easy and the distractor stimulus was highly discrepant, P300 target amplitude was larger than that elicited by the distractor stimulus, and both component types were largest over the parietal electrode sites. However, when the target/standard discrimination was difficult and the distractor stimulus was highly discrepant, the distractor stimulus elicited a P300 that was greater in amplitude frontally and shorter in latency than the target P300. Additional studies have found that the repeated distractor "typical" stimulus elicits a P3a component that is larger in amplitude over the frontal/central locations and shorter in latency than the target P3b components (Comerchero & Polich, 1998, 1999). These results suggest that the engagement of frontal lobe attentional mechanisms elicited by a difficult target stimulus detection task is a defining aspect of the stimulus context that contributes to P3a generation.

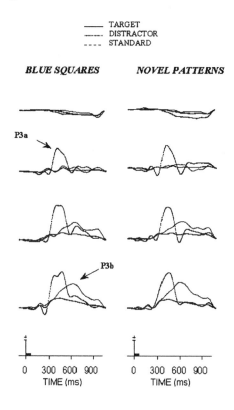

Figure 2. Grand average ERPs (n=12) from different three-stimulus oddball stimulus conditions. Subjects to respond to a target stimulus 5.5 cm diameter target circle and do not respond to a standard stimulus 5.0 cm diameter circle or to the distractor stimuli. The distractor stimuli were 23.0 cm wide squares that were all blue and always the same or different color novel patterns, with the two distractor stimulus types presented in separate conditions.

This hypothesis has been assessed systematically by comparing "typical" with "novel" distractor stimuli under easy vs. difficult target/standard discrimination tasks. Figure 2 provides an illustration of the critical ERP results from a study that compared several distractor types in a visual three-stimulus task (Demiralp et al., 2001; Polich & Comerchero, 2002). The distractor stimuli were either a blue square that was the same for each trial or variegated colored patterns that changed on each trial, with both distractor stimuli much larger than the target and standard. The task was to discriminate a 5.0 cm diameter target stimulus circle from a 4.5 cm standard stimulus circle, with the P3a subcomponent elicited by the distractor stimulus types (23.0 cm^2). The blue square and novel patterns produced P3a and P3b components that were virtually identical, such that the P3a components from the large squares were remarkably similar to those previously reported for "novel" stimuli (Courchesne et al., 1975; Simons et al., 2001; Squires et al., 1975). It is therefore reasonable to suppose that stimulus context—the relative perceptual distinctiveness among stimuli—determines both distractor and target P300 amplitude since each stimulus type produces distinct scalp topographic distributions.

2.   P300 THEORY

2.1   Context Updating

P300 amplitude is thought to index brain activity that is "required in the maintenance of working memory" when the mental model of the stimulus context is updated (Donchin et al., 1986, p. 256). Figure 3 illustrates this theoretical perspective and schematically portrays the updating processes hypothesized to produce the canonical P300 during oddball task performance. After initial sensory stimulus processing, a memory comparison evaluation is executed in which the current stimulus of the oddball sequence is compared to the previous stimulus. If no change in stimulus attributes is detected, the old "schema" or neural model of the stimulus environment is maintained, and sensory evoked potentials are recorded. However, if a new stimulus is processed the system engages attentional mechanisms to "update" the neural representation of the stimulus context and the P300 (P3b) is elicited, a process that is thought to index the ensuing memory storage operations, as P300 amplitudes are related to memory for previous stimulus presentations (Fabiani et al., 1990; Johnson, 1995; Paller et al., 1988a). A variety of cognitive factors have been delineated in support of this view, with information content, stimulus probability structure, task relevance/difficulty, and stimulus properties all found to affect P300 measures (Donchin & Coles, 1988; Johnson, 1988b; Verleger, 1988).

MEMORY COMPARISON   **NEURAL MODEL**

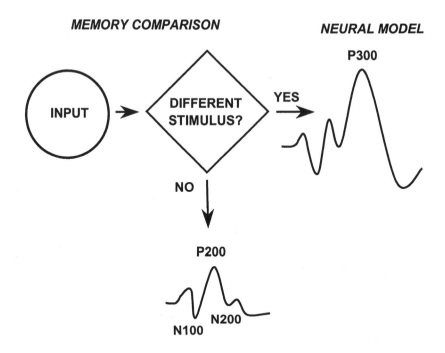

Figure 3. Schematic illustration of the context-updating model for P300 theory. Stimuli enter the processing system and a memory comparison process is engaged that ascertains whether the current stimulus is either the same as the previous stimulus or not (e.g., a standard or a target stimulus in the oddball task). If the incoming stimulus is the same, the neural model of the stimulus environment is unchanged, and signal averaging of the EEG reveals sensory evoked potentials (N100, P200, N200). If the incoming stimulus is not the same and the subject discriminates the target from the preceding standard stimulus, the neural model of the stimulus environment is changed or "updated", such that a P300 (P3b) potential is generated in addition to the sensory evoked potentials.

P300 latency is considered to be a measure of stimulus classification speed unrelated to response selection processes (Kutas et al., 1977; McCarthy & Donchin, 1981; Pfefferbaum et al., 1986), such that its timing is independent of behavioral reaction time (Duncan-Johnson, 1981; Ilan & Polich, 1999; Verleger, 1997). Given that P300 latency is an index of the processing time that occurs before response generation, it provides a temporal measure of the neural activity underlying the processes of attention allocation and immediate memory. Further, component timing is negatively correlated with processing efficiency in normal subjects: Shorter latencies are associated with superior cognitive performance from neuropsychological tests that assess how rapidly attentional resources are allocated for memory processing (e.g., Houlihan et al., 1998; Polich et al., 1983, 1990b; Polich &

Martin, 1992; Reinvang, 1999; Stelmack & Houlihan, 1994). This association is also found in clinical studies that indicate P300 latency increases as mental capability is compromised by dementing illness (e.g., Homberg et al., 1986; O'Donnell et al., 1992; Polich et al., 1986, 1990a; Potter & Barrett, 1999).

2.2   Attentional Resource Allocation

P300 is derived from neural activity such that it is necessarily affected by the physical state of its underlying physiology and therefore reflects arousal level. This interaction occurs in two ways: (1) a general arousing effect and (2) a specific or idiosyncratic effect that contributes to a complex pattern of activation that modulates information processing (Kok, 1990; Pribram & McGuiness, 1975). Arousal's tonic changes usually involve time periods on the order of minutes or hours and are manifestations of relatively slow fluctuations in the general or non-specific background arousal state of the individual, whereas phasic responses reflect the organism's energetic reaction to specific stimulus events. In this framework, tonic and phasic arousal effects originate from situational or spontaneous factors and affect the cognitive operations of attention and memory updating—i.e., the same processes hypothesized to underlie P300 generation (Polich & Kok, 1995). This theoretical interpretation is consonant with the context-updating view of P300 generation but additionally specifies a more general explanatory mechanism for the cognitive variables that affect this ERP component (cf. Donchin et al., 1986; Hillyard & Picton, 1987; Johnson, 1986).

3.   NEUROPSYCHOLOGY OF P300

3.1   P300 and the Hippocampal Formation

The precise neural origins and, therefore, the neuropsychological meaning of the P300 are as yet unknown although appreciable progress has been made in the last 20 years. Given the theoretical association of attentional and memory operations with P300, the first human studies on the neural origins of this ERP focused on the hippocampal formation. Initial reports employed depth electrodes that were implanted to help identify sources of epileptic foci in neurological patients. These recordings suggested that at least some portion of the P300 (P3b) is generated in the hippocampal areas of the medial temporal lobe (Halgren et al., 1980; McCarthy et al., 1989). However, subsequent investigations using scalp recordings on individuals after temporal lobectomy (Johnson, 1988a; Smith & Halgren,

1989), experimental excisions in monkeys (Paller et al., 1988b, 1992), and patients with severe medial temporal lobe damage (Onofrj et al., 1992; Rugg et al., 1991) found that the hippocampal formation does not contribute directly to the generation of P300 (Molnár, 1994).

3.2   P3a Neural Substrates

As outlined above, the P3a subcomponent is produced when the attentional focus required for the primary discrimination task is interrupted by an infrequent nontarget stimulus event, which does not have to be perceptually novel (Comerchero & Polich, 1999; Polich & Comerchero, 2002). ERP studies on humans with frontal lobe lesions have demonstrated that P3a requires frontal lobe function (Knight, 1984). P3a from the novel distractor stimulus for the controls evinced frontal/central maximum amplitude, whereas P3b from the target stimulus produced a parietal maximum topographic scalp distribution. However, the frontal lesion patients demonstrated a clear diminution of the P3a subcomponent for the distractor stimulus, and the usual parietal maximum for the P3b from the target stimulus. These results imply that frontal lobe engagement is necessary for P3a generation and contributes to the larger role of these mechanisms in attentional control (Knight, 1990, Knight et al., 1995).

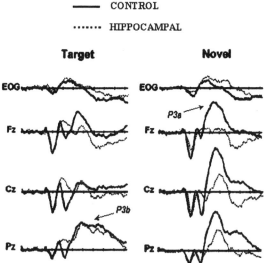

Figure 4. Grand average auditory target and novel stimulus ERPs from normal controls and bilateral hippocampal lesion patients (n=7/group). Controls demonstrate robust P3a and P3b components, whereas hippocampal patients demonstrate highly reduced P3a components over the frontal/central recording site (after Knight, 1996).

Frontal lobe activity is not the only neural source for the P3a, as the hippocampal formation has been associated with the ERP processing of "novelty" in patients with focal hippocampal lesions (Knight, 1996). Figure 4 illustrates the primary findings. P3a amplitude from novel auditory distractor stimuli for the controls yields the typical frontal/central maximum scalp topography, whereas for the patients this subcomponent is virtually eliminated over frontal electrode sites. P3b amplitude from the target stimulus is generally similar between the groups at the parietal site as observed previously. Thus, P3a generation appears to require frontal lobe attentional mechanisms and hippocampal processes driven by novelty information processing (Knight, 1997).

3.3   Frontal-to-Parietal Lobe Interactions

Given this background, possible neuropsychological mechanisms for P3a and P3b generation can be developed. Figure 5 presents a schematic model. Discrimination between target and standard stimuli in an oddball paradigm is hypothesized to initiate frontal lobe activation that reflects the attentional focus required by task performance (Pardo et al., 1991; Posner, 1992), with ERP and neuroimaging findings demonstrating frontal lobe engagement for the detection of rare or alerting stimuli (McCarthy et al., 1997; Potts et al., 1996; Verbaten et al., 1997). P3a is related to the neural changes in the anterior cingulate when incoming stimuli replace the contents of working memory, and communication of this representational change is transmitted to infero-temporal lobe stimulus maintenance mechanisms (Desimone et al., 1995). P3b reflects the operation of memory storage operations that are then initiated in the hippocampal formation with the updated output transmitted to parietal cortex (Knight, 1996; Squire & Kandell, 1999). Although the exact pathways are not yet clear (Halgren et al., 1995a, 1995bb), a variety of evidence suggests that the hippocampal formation contributes to these events, even though it is not necessary for P3b generation (Johnson, 1988a; Polich & Squire, 1993). In sum, when a distracting stimulus commands frontal lobe attention a P3a is produced; when attentional resources are allocated for subsequent memory updating after stimulus evaluation, a P3b is produced to establish the connection with storage areas in associational cortex.

As the model suggests, the neuroelectric events that underlie P300 generation stem from the interaction between frontal lobe and hippocampal/temporal-parietal function as outlined above (cf. Kirino et al., 2000; Knight, 1996). ERP and fMRI studies using oddball tasks have obtained patterns consistent with this frontal-to-temporal and parietal lobe activation pattern (He et al., 2001; Kiehl et al., 2001; Mecklinger et al.,

1998; Opitz et al., 1999). Further support comes from magnetic resonance imaging (MRI) of gray matter volumes and indicates that individual variation in P3a amplitude from distractor stimuli is correlated with frontal lobe area size, whereas P3b amplitude from target stimuli is correlated with parietal area size (Ford et al., 1994)—a finding that may underlie individual variability for observing P3a and P3b subcomponents from simple oddball tasks (cf. Polich, 1988; Squires et al., 1975).

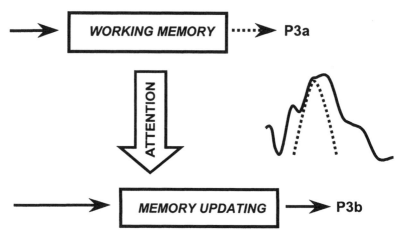

Figure 5. Schematic model of cognitive P300 activity. Sensory input is processed in parallel streams, with frontal lobe activation from attention-driven working memory changes producing P3a and temporal/parietal lobe activation from memory updating operations producing P3b. See text for explanation.

It also should be noted that the initial neural activation during auditory discrimination appears to originate from right frontal cortex (Polich et al., 1997), and that P300 amplitude is larger over the right compared to left frontal/central areas (Alexander et al., 1995; Mertens & Polich, 1997). Hence, after initial frontal processing of the incoming stimulus, activity is propagated between the cerebral hemispheres across the corpus callosum (Barcelo et al., 2000; Baudena et al., 1995; Satomi et al., 1995). This hypothesis is supported by evidence that larger callosal fiber tracts are associated with greater P300 amplitudes and shorter latencies (Alexander & Polich, 1997; Polich & Hoffman, 1998), most likely because of increased inter-hemispheric communication (cf. Driesen & Raz, 1995; Witelson, 1992). Thus, the P3a and P3b are distinct ERP components that arise from the interaction between frontal lobe attentional control over the contents of working memory and the subsequent long-term storage operations.

4. CONCLUSIONS

This overview has attempted to summarize the mechanisms underlying P3a and P3b generation. As these brain potentials are related to fundamental aspects of mental function, they offer significant promise as a means to assess normative and impaired cognitive capability. Further assessment of their neuropsychological foundations will provide additional insight into the meaning of P300. The theoretical and methodological approaches outlined here are an attempt to provide a basis for this development and are derived from contemporary research findings on factors that govern P3a and P3b production.

ACKNOWLEDGMENTS

This work was supported by NIDA Grant RO1 DA08363-05.
Figure 4 reprinted from Knight, R.T. (1996). Contribution of human hippocampal region to novelty detection. *Nature*, 383, 256-259. Copyright (2002) with permission from Macmillan Magazines Limited.

REFERENCES

Alexander, J., & Polich, J. (1995). P300 differences between sinistrals and dextrals. *Cognitive Brain Research, 2,* 277-282.

Alexander, J.E., Porjesz, B., Bauer, L.O., Kuperman, S., Mozarati, S., O'Connor, S.J. Rohrbaugh, J., Begleiter, H., & Polich, J. (1995). P300 amplitude hemispheric asymmetries from a visual discrimination task. *Psychophysiology, 32,* 467-475.

Barcelo, F., Suwazono, S., & Knight, R.T. (2000). Prefrontal modulation of visual processing in humans. *Nature Neuroscience, 3,* 399-403.

Bashore, T.R., & van der Molen, M. (1991). Discovery of P300: A tribute. *Biological Psychology, 32,* 155-171.

Baudena, P., Halgren, E., Heit, G., & Clarke, J.M. (1995). Intracerebal potentials to rare target and distractor auditory and visual stimuli. III. Frontal cortex. *Electroencephalography and Clinical Neurophysiology, 94,* 251-64.

Comerchero, M.D., & Polich, J. (1998). P3a, perceptual distinctiveness, and stimulus modality. *Cognitive Brain Research, 7,* 41-48.

Comerchero, M.D., & Polich, J. (1999). P3a and P3b from typical auditory and visual stimuli. *Clinical Neurophysiology, 110,* 24-30.

Courchesne, E. (1978). Changes in P3 waves with event repetition: Long-term effects on scalp distribution and amplitude. *Electroencephalography and Clinical Neurophysiology, 45,* 754-766.

Courchesne, E., Courchesne, R.Y., & Hillyard, S.A. (1978). The effect of stimulus deviation on P3 waves to easily recognized stimuli. *Neuropsychologia, 16,* 189-199.

Courchesne, E., Hillyard, S.A., & Galambos, R. (1975). Stimulus novelty, task relevance and the visual evoked potential in man. *Electroencephalography and Clinical Neurophysiology, 39,* 131-143.

Courchesne, E., Kilman, B.A., Galambos, R., & Lincoln, A. (1984). Autism: Processing of novel auditory information assessed by event-related brain potentials. *Electroencephalography and Clinical Neurophysiology, 59,* 238-248.

Demiralp, T., Ademoglu, A., Comerchero, M., & Polich, J. (2001). Wavelet analysis of P3a and P3b. *Brain Topography, 13,* 251-267.

Desimone, R., Miller, E.K., Chelazzi, L., & Lueschow, A. (1995). Multiple memory systems in the visual cortex. In M.S. Gazzaniga (Ed.), *The cognitive neurosciences* (pp. 475-486). Cambridge, MA: MIT Press.

Donchin, E., & Coles, M.G.H. (1988). Is the P300 component a manifestation of context updating? *Behavioral and Brain Sciences, 11,* 357-374.

Donchin, E., Karis, D., Bashore, T.R., Coles, M.G.H., & Gratton, G. (1986). Cognitive psychophysiology and human information processing. In M.G.H. Coles, E. Donchin, & S.W. Porges (Eds.), *Psychophysiology: Systems, processes, and applications* (pp. 244-267). New York: The Guilford Press.

Driesen, N.R., & Raz, N. (1995). The influence of sex, age, and handedness of corpus callosum morphology: A meta-analysis. *Psychobiology, 23,* 240-247.

Duncan-Johnson, C.C. (1981). P300 latency: A new metric of information processing. *Psychophysiology, 18,* 207-215.

Emmerson, R., Dustman, R., Shearer, D., & Turner, C. (1990). P300 latency and symbol digit correlations in aging. *Experimental Aging Research, 15,* 151-159.

Fabiani, M., Karis, D., & Donchin, E. (1990). Effects of mnemonic strategy manipulation in a von Restorff paradigm. *Electroencephalography and Clinical Neurophysiology, 75,* 22-35.

Fabiani, M, Friedman, D., & Ching, J. (1998). Individual differences in P3 scalp distribution in older adults, and their relationship to frontal lobe function. *Psychophysiology, 35,* 698-708.

Falkenstein, M., Koshlykova, N.A., Kiroj, V.N., Hoormann, J., & Hohnsbein, J. (1995). Late ERP components in visual and auditory go/nogo tasks. *Electroencephalography and Clinical Neurophysiology, 96,* 36-43.

Ford, J.M., Sullivan, E., Marsh, L., White, P., Lim, K., & Pfefferbaum, A. (1994). The relationship between P300 amplitude and regional gray matter volumes depends on the attentional system engaged. *Electroencephalography and Clinical Neurophysiology, 90,* 214-28.

Friedman, D., & Simpson, G. (1994). ERP amplitude and scalp distribution to target and novel events: Effects of temporal order in young, middle-age, and older adults. *Cognitive Brain Research, 2,* 49-63.

Friedman, D., Kazmerski, V.A., & Cycowicz, Y.M. (1998). Effects of aging on the novelty P3 during attend and ignore oddball tasks. *Psychophysiology, 35,* 508-520.

Grillon, C., Courchesne, E., Ameli, R., Elmasian, R., & Braff, D. (1990). Effects of rare non-target stimuli on brain electrophysiological activity and performance. *International Journal of Psychophysiology, 9,* 257-267.

Halgren, E., Squires, N.K., Wilson, C., Rohrbaugh, J., Babb, T., & Crandall, P. (1980). Endogenous potentials in the human hippocampal formation and amygdala by infrequent events. *Science, 210,* 803-805.

Halgren, E., Baudena, P., Clarke, J.M., Heit, G., Liegeois, C., Chauvel, P., & Musolino, A. (1995a). Intracerebral potentials to rare target and distractor auditory and visual stimuli. I. Superior temporal plane and parietal lobe. *Electroencephalography and Clinical Neurophysiology, 94,* 191-220.

Halgren, E., Baudena, P., Clarke, J. M., Heit, G., Marinkovic, K., Devaux, B., Vignal, J.P., & Biraben, A. (1995b). Intracerebral potentials to rare target and distractor auditory and visual stimuli. II. Medial, lateral and posterior temporal lobe. *Electroencephalography and Clinical Neurophysiology, 94*, 229-250.

He, B., Lian, J., Spencer, K.M., Dien, J., & Donchin, E. (2001). A cortical potential imaging analysis of the P300 and novelty P3 components. *Human Brain Mapping, 12*, 120-130.

Hillyard, S.A., & Picton, T.W. (1987). Electrophysiology of cognition. In F. Plum (Ed.), *Handbook of physiology* (pp. 519-584.). Baltimore: American Physiological Society.

Homberg, V., Hefter, H., Granseyer, G., Strauss, W., Lange, H., & Hennerici, M. (1986). Event-related potentials in patients with Huntington's disease and relatives at risk in relation to detailed psychometry. *Electroencephalography and Clinical Neurophysiology, 63*, 552-569.

Houlihan, M., Stelmack, R., & Campbell, K. (1998). P300 and cognitive ability: Assessing the roles of processing speed, perceptual processing demands and task difficulty. *Intelligence, 26*, 9-25.

Ilan, A.B., & Polich, J. (1999). P300 and response time from a manual Stroop task. *Clinical Neurophysiology, 110*, 367-373.

Johnson, R. (1986). A triarchic model of P300 amplitude. *Psychophysiology, 23*, 367-384.

Johnson, R. (1988a). Scalp-recorded P300 activity in patients following unilateral temporal lobectomy. *Brain, 111*, 1517-29.

Johnson, R. (1988b). The amplitude of the P300 component of the event-related potential: Review and synthesis. In P. Ackles, J.R. Jennings, & M.G.H. Coles (Eds.), *Advances in psychophysiology: A research annual, Vol. 3* (pp. 69-137). Greenwich, CT: JAI Press, Inc.

Johnson, R. (1995). Event-related potential insights into the neurobiology of memory systems. In F. Boller & J. Grafman (Eds.), *Handbook of neuropsychology, Vol. 10* (pp. 135-63). Elsevier: Amsterdam.

Katayama, J., & Polich, J. (1996a). P300 from one-, two-, and three-stimulus auditory paradigms. *International Journal of Psychophysiology, 23*, 33-40.

Katayama, J., & Polich, J. (1996b). P300, probability, and the three-tone paradigm. *Electroencephalography and Clinical Neurophysiology, 100*, 555-562.

Katayama, J., & Polich, J. (1998). Stimulus context determines P3a and P3b. *Psychophysiology, 35*, 23-33.

Kiehl, K.A., Laurens, K.R., Duty, T.L., Forster, B.B., & Liddle, P.F. (2001). Neural sources involved in auditory target detection and novelty processing: An event-related fMRI study. *Psychophysiology, 38*, 133-142.

Kirino, E., Belger, A., Goldman-Rakic, P., & McCarthy, G. (2000). Prefrontal activation evoked by infrequent target and novel stimuli in a visual target detection task: An event-related functional magnetic resonance study. *The Journal of Neuroscience, 20*, 6612-6618.

Knight, R.T. (1984). Decreased response to novel stimuli after prefrontal lesions in man. *Electroencephalography and Clinical Neurophysiology, 59*, 9-20.

Knight, R.T. (1990). Neural mechanisms of event-related potentials from human lesion studies. In J. Rohrbaugh, R. Parasuraman, & R. Johnson (Eds). *Event-related brain potentials: Basic issues and applications* (pp. 3-18). New York: Oxford Press.

Knight, R.T. (1996). Contribution of human hippocampal region to novelty detection. *Nature, 383*, 256-259.

Knight, R.T. (1997). Distributed cortical network for visual attention. *Journal of Cognitive Neuroscience, 9*, 75-91.

Knight, R., Grabowecky, M., & Scabini, D. (1995). Role of human prefrontal cortex in attention control. *Advances in Neurology, 66*, 21-34.

Kok, A. (1990). Internal and external control: A two-factor model of amplitude change of event-related potentials. *Biological Psychology, 74*, 203-236.

Kutas, M., McCarthy, G., & Donchin, E. (1977). Augmenting mental chronometry: The P300 as a measure of stimulus evaluation. *Science, 197*, 792-795.

McCarthy, G., & Donchin, E. (1981). A metric for thought: A comparison of P300 latency and reaction time. *Science, 211*, 77-80.

McCarthy, G., Wood, C.C., Williamson, P.D., & Spencer, D. (1989). Task-dependent field potentials in human hippocampal formation. *Journal of Neuroscience, 9*, 4235-4268.

McCarthy, G., Luby, M., Gore, J., & Goldman-Rakic, P. (1997). Infrequent events transiently activate human prefrontal and parietal cortex as measured by functional MRI. *Journal of Neurophysiology, 77*, 1630-1634.

Mecklinger, A., Maess, B., Opitz, B., Pfeifer, E., Cheyne, D., & Weinberg, H. (1998). A MEG analysis of the P300 in visual discrimination tasks. *Electroencephalography and Clinical Neurophysiology, 108*, 45-56.

Mertens, R., & Polich, J. (1997). P300 hemispheric differences from oddball, verbal, and spatial tasks. *Psychophysiology* (abstract), *34*, S64.

Molnár, M. (1994). On the origin of the P300 event-related potential component. *International Journal of Psychophysiology, 17*, 129-44.

O'Donnell, B.F., Friedman, S., Swearer, J.M., & Drachman, D. (1992). Active and passive P300 latency and psychometric performance: influence of age and individual differences. *International Journal of Psychophysiology, 12*, 187-195.

Opitz, B., Mecklinger, A., Von Cramon, D.Y., & Kruggel, F. (1999). Combining electrophysiological and hemodynamic measures of the auditory oddball. *Psychophysiology, 36*, 142-147.

Onofrj, M., Fulgente, T., Nobiolio, D., Malatesta, G., Bazzano, S., Colamartino, P., & Gambi, D. (1992). P300 recordings in patients with bilateral temporal lobe lesions. *Neurology, 42*, 1762-7.

Paller, K.A., McCarthy, G., Roessler, E., Allison, T., & Wood, C.C. (1992). Potentials evoked in human and monkey medial temporal lobe during auditory and visual oddball paradigms. *Electroencephalography and Clinical Neurophysiology, 84*, 269-279.

Paller, K.A., McCarthy, G., & Wood, C.C. (1988a). ERPs predictive of subsequent recall and recognition performance. *Biological Psychology, 26*, 269-276.

Paller, K.A., Zola-Morgan, S., Squire, L.R., & Hillyard, S.A. (1988b). P3-like brain waves in normal monkeys and in monkeys with medial temporal lesions. *Behavioral Neuroscience, 102*, 714-725.

Pardo, J.V., Fox, P., & Raichle, M. (1991). Localization of human system for sustained attention by positron emission tomography. *Nature, 349*, 61-64

Pfefferbaum, A., Ford, J.M., Weller, B.J., & Kopell, B.S. (1985). ERPs to response production and inhibition. *Electroencephalography and Clinical Neurophysiology, 60*, 423-434.

Pfefferbaum, A., & Ford, J.M. (1988). ERPs to stimuli requiring response production and inhibition: effects of age, probability and visual noise. *Electroencephalography and Clinical Neurophysiology, 71*, 55-63.

Pfefferbaum, A., Christensen, C., Ford, J.M., & Kopell, B.S. (1986). Apparent response incompatibility effects on P300 latency depend on the task. *Electroencephalography and Clinical Neurophysiology, 64*, 424-437.

Picton, T.W. (1992). The P300 wave of the human event-related potential. *Journal of Clinical Neurophysiology, 9*, 456-479.

Polich, J. (1988). Bifurcated P300 peaks: P3a and P3b revisited? *Journal of Clinical Neurophysiology, 5*, 287-294.

Polich, J. (1999). P300 in clinical applications. In E. Niedermeyer & F. Lopes da Silva (Eds.), *Electroencephalography: Basic principles, clinical applications and related fields* (4th ed., pp. 1073-1091). Baltimore-Munich: Urban & Schwarzenberg.

Polich, J., & Comerchero, M.D. (2003). P3a and P3b from novel and typical stimuli. Submitted.

Polich, J. & Hoffman, L.D. (1998). P300 and handedness: On the possible contribution of corpus callosal size to ERPs. *Psychophysiology, 35*, 497-507.

Polich, J., & Kok, A. (1995). Cognitive and biological determinants of P300: An integrative review. *Biological Psychology, 41*, 103-146.

Polich, J., & Martin, S. (1992). P300, cognitive capability, and personality: A correlational study of university undergraduates. *Personality and Individual Differences, 13*, 533-543.

Polich, J., & Squire, L. (1993). P300 from amnesic patients with bilateral hippocampal lesions. *Electroencephalography and Clinical Neurophysiology, 86*, 408-417.

Polich, J., Howard, L., & Starr, A. (1983). P300 latency correlates with digit span. *Psychophysiology, 20*, 665-669.

Polich, J., Ladish, C., & Bloom, F.E. (1990a). P300 assessment of early Alzheimer's disease. *Electroencephalography and Clinical Neurophysiology, 77*, 179-189.

Polich, J., Ladish, C., & Burns, T. (1990b). Normal variation of P300 in children: Age, memory span, and head size. *International Journal of Psychophysiology, 9*, 237-248.

Polich, J., Ehlers, C.L., Otis, S., Mandell, A.J., & Bloom, F.E. (1986). P300 latency reflects cognitive decline in dementing illness. *Electroencephalography and Clinical Neurophysiology, 63*, 138-14.

Polich, J., Alexander, J.E., Bauer, L.O., Kuperman, S., Rohrbaugh, J., Mozarati, S., O'Connor, S.J., Porjesz, B., & Begleiter, H. (1997). P300 topography of amplitude/latency correlations. *Brain Topography, 9*, 275-282.

Posner, M.I. (1992). Attention as a cognitive neural system. *Current Directions in Psychological Science, 1*, 11-14.

Potts, G.F., Liotti, M., Tucker, D.M., & Posner, M.I. (1996). Frontal and inferior temporal cortical activity in visual target detection: Evidence from high spatially sampled event-related potentials. *Brain Topopography, 9*, 3-14.

Potter, D.D., & Barrett, K. (1999). Assessment of mild head injury with ERPs and neuropsychological tasks. *Journal of Psychophysiology, 13*, 173-189.

Pribram, K.H., & McGuinness, D. (1975). Arousal, activation, and effort in the control of attention. *Psychological Review, 82*, 116-149.

Reinvang, I. (1999). Cognitive event-related potentials in neuropsychological assessment. *Neuropsychology Review, 9*, 231-248.

Rugg, M.D., Pickles, C., Potter, D., & Roberts, R. (1991). Normal P300 following extensive damage to the left medial temporal lobe. *Journal of Neurology, Neurosurgery, and Psychiatry, 54*, 217-222.

Satomi, K., Horai, T., Kinoshita, Y., & Wakazono, A. (1995). Hemispheric asymmetry of event-related potentials in a patient with callosal disconnection syndrome: A comparison of auditory, visual, and somatosensory modalities. *Electroencephalography and Clinical Neurophysiology, 94*, 440-9.

Simons, R.F., Graham, F.K., Miles, M.A., & Chen, X. (2001). On the relationship of P3a and the novelty-P3. *Biological Psychology, 56*, 207-218.

Smith, M.E., & Halgren, E. (1989). Dissociation of recognition memory components following temporal lobe lesions. *Journal of Experimental Psychology: General, 15*, 50-60.

Spencer, K.M., Dien, J., & Donchin, E. (1999). A componental analysis of the ERP elicited by novel events using a dense electrode array. *Psychophysiology, 36*, 409-414.

Squire, L.R., & Kandel, E.R. (1999). *Memory from mind to molecules*. New York: Scientific American Library.

Squires, N.K., Squires, K., & Hillyard, S.A. (1975). Two varieties of long-latency positive waves evoked by unpredictable auditory stimuli in man. *Electroencephalography and Clinical Neurophysiology, 38,* 387-401.

Stelmack, R.M., & Houlihan, M. (1994). Event-related potentials, personality and intelligence: Concepts, issues, and evidence. In D.H. Saklofske & M. Zaidner (Eds.), *International handbook of personality and intelligence* (pp. 349-365). New York: Plenum Press.

Sutton, S., Braren, M., Zubin, J., & John, E.R. (1965). Evoked potential correlates of stimulus uncertainty. *Science, 150,* 1187-1188.

Suwazono, S., Machado, L., & Knight, R.T. (2000). Predictive value of novel stimuli modifies visual event-related potentials and behavior. *Clinical Neurophysiology, 111,* 29-39.

Verbaten, M.N., Huyben, M.A., & Kemner, C. (1997). Processing capacity and the frontal P3. *International Journal of Psychophysiology, 25,* 237-248.

Verleger, R. (1997). On the utility of P3 latency as an index of mental chronometry. *Psychophysiology, 34,* 131-156.

Verleger, R. (1988). Event-related potentials and cognition: A critique of the context updating hypothesis and an alternative interpretation of P3. *Behavioral and Brain Sciences, 11,* 343-356.

Witelson, S.F. (1992). Cognitive neuroanatomy: A new era. *Neurology, 42,* 709-713.

Yamaguchi, S., & Knight, R. (1991a). Age effects on the P300 to novel somatosensory stimuli. *Electroencephalography and Clinical Neurophysiology, 78,* 297-301.

Yamaguchi, S., & Knight, R.T. (1991b). P300 generation by novel somatosensory stimuli. *Electroencephalography and Clinical Neurophysiology, 78,* 50-55.

Chapter 6

LATERAL AND ORBITAL PREFRONTAL CORTEX CONTRIBUTIONS TO ATTENTION

KAISA M. HARTIKAINEN AND ROBERT T. KNIGHT
Department of Psychology and Helen Wills Neuroscience Institute, University of California, Berkeley, CA, USA

1.   PREFRONTAL CORTEX

The prefrontal cortex (PFCx) can be divided into the lateral, orbital and medial PFCx, which all contribute to attentional and novelty processing and flexible behaviors. The cytoarchitecture of lateral prefrontal cortex (LPFCx) is highly organized, differentiated, distinctly layered, granular isocortex while the orbitofrontal cortex (OFCx) is structurally more heterogeneous, less differentiated, agranular limbic cortex (Barbas, 2000). The LPFCx is interconnected with parietal/occipital visual association areas, posterior parietal heteromodal areas, and inferior temporal visual association areas (Kaufer & Lewis, 1999). Other main circuitries of the LPFCx include reciprocal connections with the cingulate and the orbitofrontal cortex. In contrast, the OFCx has extensive direct and indirect connections with limbic areas such as the amygdala complex, hypothalamus, and the hippocampal formation (Cavada et al., 2000), with additional interactions to the inferior temporal visual association areas and LPFCx (Kaufer & Lewis, 1999). The main connections of the LPFCx to association areas and the OFCx to limbic areas determine the primary behavioral functions of these areas, with the LPFCx well-suited for attentional and executive function and the OFCx for affective and reward-related functions.

The PFCx allows for departure from automated actions (Mesulam, 1986). Adjusting behavior depending on the current situation, social context, and foresight requires inhibiting responding to the most salient stimuli as well as inhibiting previously acquired responses. Thus, inhibition is an essential component of cognitive flexibility and creative behaviors that tend to be compromised in patients with PFCx damage. Response inhibition has been suggested to be a general PFCx function that operates across different cognitive processes and brain regions (Roberts & Wallis, 2000). Both the lateral and the orbital PFCx perform general inhibitory functions, but the distinct cognitive processes that are modulated by these cortical areas differ and reflect the distinct neural circuitries in which they are imbedded. Dias et al. (1996, 1997) have suggested that the lateral prefrontal cortex is responsible for inhibitory control of attentional selection, while the orbitofrontal cortex is responsible for inhibitory control of affective responses. Hence, damage to the lateral prefrontal cortex leads to impairment in shifting attention from one perceptual dimension to another while damage to orbitofrontal cortex leads to an inability to alter behavior when the emotional significance of the stimuli change (Dias et al., 1996, 1997). Deficits in inhibitory mechanisms lead to different clinical symptoms in patients with lateral and orbital PFCx damage. Lack of inhibition is a likely explanation for impulsivity, socially inappropriate or the disinhibited behaviors often observed after orbitofrontal damage (Levine et al., 1999), whereas deficits in attention, inflexible cognition, stimulus bound and perseverative behaviors may be signs of inhibitory deficits in lateral PFCx damage.

2.   ELECTROPHYSIOLOGY AND LESION METHODS

Behavioral measurements, recordings of neuronal activity of single cells and neural populations, as well as blood flow changes in response to neural activity have been used to study PFCx function in both intact and lesioned brains. Despite limitations of each method, converging evidence from different techniques provides a more reliable and richer understanding than any single approach to the roles of prefrontal cortices in cognition, emotion, and behavior. The focus of this chapter is on results obtained from electrophysiological studies on neurological patients with focal lateral or orbital prefrontal damage.

Electrophysiological techniques such as electroencephalography (EEG) and event-related brain potentials (ERPs) provide important approaches to study attention and other cognitive processes in humans (Näätänen, 1992). Lesion studies and functional magnetic resonance imaging (fMRI) help to delineate the brain regions engaged in cognitive processing. However, mental events occur so rapidly that fMRI methods are often not amenable to neuroimaging cognition (McIntosh et al., 1994; 1999). Despite improving temporal resolution of the fMRI method, the sluggishness of the hemodynamic response underlying the fMRI signal restricts the temporal information into the range of seconds. ERPs have millisecond temporal resolution and are therefore well suited for assessing the kinetics of human cognition in real time. In addition, fMRI is susceptible to artifacts originating from neighboring anatomical structures (e.g., air-filled sinuses), which can compromise reliable imaging of brain regions such as the orbitofrontal cortex. Thus, combining electrophysiology with the lesion method provides both temporal and spatial information and allows insight into the dynamics and neural circuitry of cognitive processes.

Neuroanatomical information from lesion studies (Knight et al., 1998; Knight & Scabini, 1998), intracranial recordings (Baudena et al., 1995; Halgren et al., 1995a, Halgren et al., 1995b; Halgren et al., 1998) and combined neuroimaging and ERP studies (Heinze et al., 1994; Opitz et al., 1999a; Opitz et al., 1999b) have delineated the neural regions responsible for generating several widely studied cognitive ERP components. For instance, attention sensitive visual ERPs, including a positive (P1, 110-160 milliseconds) and a subsequent negative potential (N1, 125-225 milliseconds) have been localized to the extrastriate cortex (Gonzalez et al., 1994; Hillyard & Anllo-Vento, 1998; Martinez et al., 1999), and fMRI studies have confirmed extrastriate attention modulation (Brefczynski & DeYoe, 1999; Chawla et al., 1999; Kastner et al., 1999). Electrophysiological and neurological techniques have also defined a distributed cortical-limbic network activated within 150-400 milliseconds after a novel irrelevant stimulus event (Alain et al., 1998; Halgren et al., 1998, Knight, 1984). Novel stimuli generate the P3a ERP, which is a positive-going component that occurs at about 300-400 milliseconds and is maximal over the anterior scalp. This novelty ERP is proposed to be a central marker of the orienting response (Bahramali et al., 1997; Courchesne et al., 1975; Escera et al., 1998; Knight, 1984; Yamaguchi & Knight, 1991). ERP evidence derived from neurological patients and intracranial ERP recordings in pre-surgical epileptics has revealed that a distributed neural network including the lateral and orbital PFCx, hippocampal formation, anterior cingulate and temporal–parietal cortex is involved in detecting and encoding novel information (Halgren et al., 1998; Knight 1996; Knight 1997; Knight & Scabini, 1998; Verleger et al., 1994; Yamaguchi & Knight, 1991; Yamaguchi & Knight, 1992). Neuroimaging has provided confirmation on the neuroanatomy of this novelty processing system, which engages involuntary attention (Clark et al., 2000; Downar et al., 2000; McCarthy et al., 1997; Menon et al., 1997; Opitz et al., 1999a; Opitz et al., 1999b; Stern et al., 1996; Tulving et al., 1994; Tulving et al., 1996; for a review see Friedman et al., 2001).

Strong convergence of lesion/ERP/fMRI data also has been obtained in voluntary attention paradigms. Voluntary stimulus detection generates a classic P300 or P3b potential that occurs between 300 to 700 milliseconds, has a posterior scalp maxima, and is primarily sensitive to attentional and cognitive factors rather than the physical properties of the stimulus (for reviews, see Picton, 1992 and Polich, 1998). The P300 can be triggered by detection of auditory, visual, somatosensory, and olfactory stimuli as well as by detection of missing stimuli in a train of irrelevant stimuli. The P300 response to missing stimuli highlights the importance of cognitive factors over physical properties in the generation of these late ERP components.

P300 in a voluntary target detection paradigm is referred to as P3b to distinguish it from P3a generated by task-irrelevant novel stimuli. P3a is maximal over fronto-central scalp areas and peaks in amplitude about 50 milliseconds prior to P3b activity, which is maximal over parietal areas. Task-relevant and predictable stimuli lead to small P3a and large P3b responses, while unexpected and novel stimuli result in increased prefrontal P3a amplitude. Several explanations as to the functional significance of P300 have been offered, with most models focusing on attentional and mnemonic mechanisms. P300 amplitude depends on a variety of factors such as probability, context, and relevance of the stimuli, as well as the cognitive processes engaged by the behavioral task (Donchin & Coles, 1988; Katayama & Polich, 1998) The lack of a unitary theory on the functional significance of the P300 reflects the fact that multiple brain regions and cognitive processes generate scalp positivities between 300 to 700 milliseconds after stimulus presentation that contribute to P300. In addition, a variety of evidence indicates that the temporo-parietal junction contributes to P300 generation. More important, both P3b and P3a are attenuated by lesions in the temporo-parietal junction in all sensory modalities (Figure 1; Knight, 1997; Knight et al., 1989; Yamaguchi & Knight, 1991). Furthermore, event-related fMRI studies have confirmed temporo-parietal junction activation during voluntary event detection (Clark et al., 2000; Linden et al., 1999; McCarthy et al., 1997; Menon et al., 1997). Thus, electrophysiological methods in conjunction with lesion studies have proved to be valuable approaches for identifying the neural circuitries involved in cognitive processing.

3.   LATERAL PREFRONTAL CORTEX

The LPFCx has been implicated in multiple cognitive processes such as executive control, attention, language, and memory (Chao & Knight, 1998; Corbetta, 1998; Dronkers et al., 2000; Fuster et al., 2000; Knight et al., 1998;

McDonald et al., 2000). The crucial role of the LPFCx function in intact human cognition and behavior is indicated by the variety of neurological and psychiatric disorders linked to LPFCx dysfunctions, such as schizophrenia, depression, attention deficit disorder, stroke, Parkinson's disease and frontal lobe dementia (Akbarian et al., 1995, 1996; Jagust, 1999; Miller et al., 1991; Rosen et al., 2001; Stamm et al., 1993; Weinberger et al., 1986; 1992, Wilkins et al., 1987). Studies using fMRI have also defined the role of the LPFCx in working memory, response conflict, novelty processing, and attention (Barch et al. 2000; Botvinick et al., 1999; D'Esposito et al., 1995, 1999a; D'Esposito et al., 1999b; D'Esposito et al., 1999c; Downar et al., 2000; Jonides et al., 1993, 1998; Owen et al., 1998; Prabhakaran et al., 2000; Rypma & D'Esposito, 2000). In accordance with the imaging results, patients with lesions to the LPFCx have deficits in working memory (Harrington et al., 1998; Müller et al., 2002; Stone et al., 1998), response monitoring (Gehring & Knight, 2000), novelty processing (Knight 1984; Knight & Scabini, 1998), and attention (Barceló et al., 2000; Knight et al., 1998). Evidence combining lesion, electrophysiological and fMRI data has also been essential for delineating the different roles of LPFCx in attentional mechanisms, including early modulation of primary sensory areas, later modulation of association areas as well as involvement in the novelty-driven involuntary attention network.

Single cell recordings in monkeys (Rainer et al., 1998a, Rainer et al., 1998b), lesion studies in humans (Barceló et al., 2000; Knight, 1997; Nilesen-Bohlman & Knight, 1999) and monkeys (Rossi et al., 1999), as well as blood flow data (Büchel & Friston, 1997; Chawla et al., 1999; Corbetta, 1998; Hopfinger et al., 2000; Kastner et al., 1999; Rees et al., 1997) have linked LPFCx to early attentional modulation of extrastriate cortex. Lateral prefrontal cortex attention effects span from early modulation of extrastriate activity beginning 125 milliseconds after stimulus delivery to subsequent visual processing extending throughout the ensuing 500 milliseconds (Barceló et al., 2000).

The LPFCx exerts both facilitatory and inhibitory modulation of posterior sensory and perceptual areas and contributes to both involuntary and voluntary attention networks. Facilitatory modulation of extrastriate activity can be detected as enhancement of visual P1 and N1 potentials in the 100-200 milliseconds range (Mangun, 1995). Facilitatory PFCx modulatory effects on task relevant stimuli are not limited to early sensory processing but also include later processing stages and brain areas (Barceló et al., 2000). Even though in simple detection tasks LPFCx does not appear to have significant contribution to brain potentials (e.g., N2 and P3b) reflecting target detection (Knight & Nakada, 1998; Knight & Scabini, 1998), more demanding cognitive tasks seem to rely on LPFCx modulation of posterior association areas (Swick, 1998; Swick & Knight, 1999). In addition to facilitatory modulation of relevant stimuli, LPFCx exerts inhibitory modulation of irrelevant stimuli. Consequently, LPFCx damage leads to increased distractibility (Bartus & Levere 1977; Chao & Knight, 1995; 1998; Malmo 1942; Woods & Knight, 1986). Increased distractibility is believed to partially explain the attentional deficits after brain damage (Kaipio et al., 1999). Electrophysiological signs of increased distractibility are enhanced ERP potentials to task-irrelevant stimuli, such as primary auditory cortex evoked response amplitude increase to distractors in LPFCx patients (Chao & Knight, 1998). This inhibitory control of early sensory processing has been linked to a prefrontal-thalamic gating system (Guillery et al., 1998; Knight et al., 1998).

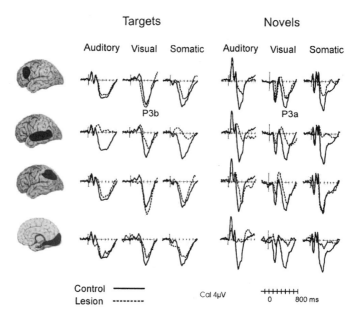

Figure 1. Grand averaged ERPs from lesion patient and control groups in a simple detection paradigm across stimulus modalities illustrate the effects of frontal, temporo-parietal junction, parietal, and hippocampal lesions on target and novelty processing. Brain images at the left illustrate the average lesion location. Prefrontal lesions reduced novelty P3a across modalities, but had no effect on the target P3b. Temporo-parietal junction lesions affect both novelty P3a and target P3b leading to marked reductions in amplitudes to auditory and somatosensory stimuli and partial reduction to visual stimuli. Lateral parietal lesion had no significant effect on either P3a or P3b amplitudes or latencies. Hippocampal damage leads to significant reductions in P3a over the frontal sites, but P3b remained intact.

In addition to its critical role in voluntary attention, the LPFCx is a key component of the novelty network that engages involuntary attentional mechanisms. Mismatch negativity (MMN) studies indicate that lateral prefrontal cortex initiates the novelty detection cascade prior to activation of other brain regions. If the novel event is sufficiently engaging, posterior cortical and medial temporal regions are recruited for further processing (Alain et al., 1998; Alho et al., 1994; Knight, 1996). PFCx novelty activation recorded with ERPs or neuroimaging habituates to repeated exposures to novel events and is modality independent (Knight, 1984; Knight & Scabini, 1998; Peterson et al., 1999; Raichle et al., 1994; Yamaguchi & Knight, 1991). Furthermore there is a marked reduction of novelty P3a in patients with LPFCx damage, whereas the P3b in a simple detection paradigm remains largely unaffected (Knight & Scabini, 1998, Figure 1; Daffner et al., 2000). These findings highlight the significant contributions of the PFCx to involuntary attention networks.

Involuntary and voluntary attentional mechanisms rely on distributed neural networks comprised of multiple brain areas. Figure 1 illustrates the value of the lesion method in determining contributions of specific brain areas to attentional and novelty processing. Similar to LPFCx, lesions of the hippocampal formation lead to clear P3a amplitude reduction, but have no significant effect on P3b (Knight, 1996). This finding provides evidence for the role of the hippocampal formation in novelty detection. Contrary to its involvement in novelty P3a generation, the hippocampal formation does not seem to contribute significantly to most scalp recorded target P3bs. In contrast, lesions of the temporo-parietal area lead to reduction in both P3a and P3b in all modalities (Knight, 1997; Knight et al., 1989; Verleger et al. 1994; Yamaguchi & Knight, 1991). Hence, the temporo-parietal junction is critical in multimodal processes involving both irrelevant novel and relevant recurring events engaging both involuntary and voluntary attentional mechanisms. Lateral parietal lesions can affect either P3a or P3b. Lateral parietal lesions serve as a brain-damage control comparison, since the ERP amplitude reductions from focal brain damage to the LPFCx, temporo-parietal junction, and hippocampal formation are not the result of general brain lesion effects, but rather are specific to lesion location and disrupting the circuits involved in novelty and target processing.

4. ORBITAL PREFRONTAL CORTEX

The extensive neuroanatomical connections of the orbital prefrontal cortex with the limbic system (Barbas, 2000; Cavada, 2000; Price, 1999), as well as its connections to lateral prefrontal cortex make it a region well suited for integrating emotion and motivation with cognition and behavior. The orbitofrontal cortex is believed to play a variety of roles in guiding adaptive, motivated, and emotion regulated behaviors. Clinical evidence since the landmark case of Phineas Gage in 1848 has highlighted the significance of the orbitofrontal cortex in emotional and social behavior (Dimitrov et al., 1999; Eslinger, 1999; Harlow, 1993; Macmillan, 2000; Nies, 1999). Lesions of the orbitofrontal cortex result in impaired social skills, emotional lability, and decreased impulse control.

In contrast to well-preserved cognitive skills and only subtle difficulties in formal neuropsychological tests, the effect of orbitofrontal damage on a patient's social behavior is considerable and may cause significant adverse consequences in their personal lives. Several reasons underlie impaired social skills in orbitofrontal patients: impaired insight (Leduc et al., 1999), difficulty in inferring mental states of others (Stone et al., 1998), failure to use emotions in guiding decisions (Bechara et al., 2000; Damasio, 1996), deficits in emotion recognition (Hornak et al., 1996), and impaired knowledge of moral rules or an inability to apply them (Anderson et al., 1999). Further functions assigned to the orbitofrontal cortex that may be crucial for successful social behavior include labeling reward values to outcomes of voluntary action and updating the reward contingencies in a rapidly changing environment (Rolls, 2000), inhibiting previously but no longer rewarded behaviors (Dias et al., 1996, 1997), and modulating orientation to irrelevant environmental stimuli (Rule et al., in press).

Although patients with orbitofrontal lesions often present with emotional lability and impulsive behavior, advanced dorsolateral prefrontal cortex lesions typically result in blunted affect and apathy. The blunted emotional state seen in dorsolateral patients is accompanied by attenuation of electrophysiological responses to novel stimuli (Paradiso et al., 1999), reflecting impairment in involuntary attentional mechanisms (Knight, 1984, Figure 1). In contrast, orbitofrontal patients show electrophysiological enhancement to novel environmental sounds (Rule et al., in press) as well as context dependent enhancement of responses to recurring events (Hartikainen et al., 2001, Figure 2). These enhanced electrophysiological responses suggest a failure in inhibitory mechanisms that may also underlie the impulsive behavior observed in OFCx patients.

In contrast to diminished late ERP amplitudes seen in association with brain damage to LPFCx, temporo-parietal junction, and hippocampal formation, significantly enhanced ERPs are observed subsequent to OFCx damage. These enhanced ERP responses provide electrophysiological evidence for the inhibitory role of OFCx in humans (Hartikainen et al., 2001; Rule, 2000). Unlike in LPFCx damage, where the impairment of inhibitory control can be observed using traditional neuropsychological testing such as the Wisconsin card sorting test, in OFCx damage many neuropsychological "frontal lobe" measures remain intact (Stuss at al., 2000). Thus, neuropsychological testing often fails to detect deficits in the inhibitory control in OFCx patients, despite sometimes significant adverse effects on everyday life due to disinhibition, whereas enhanced ERPs show promise for laboratory detection of neural disinhibition in OFCx patient's responses (Hartikainen et al., 2001; Rule et al., in press).

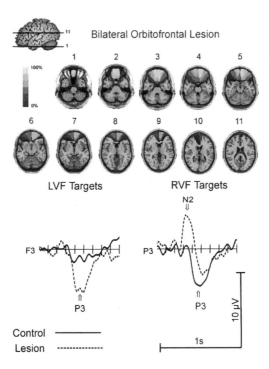

Figure 2. Enhanced ERPs to visual targets subsequent to bilateral orbitofrontal damage. Lesion reconstruction of 6 bilateral orbitofrontal patients is shown in the upper part of the figure. Average lesion location is indicated with shades of gray in the orbitofrontal area. The gray scale corresponds to lesion overlap across patients. The orbitofrontal lesions included bilateral damage in areas 10, 11, 12, and 13 with maximal damage in ventromedial orbitofrontal cortex. Orbitofrontal patients and age-matched controls discriminated between upright and inverted triangles (target). Targets were randomly presented in the left (LVF) or right visual (RVF) hemifield (150 milliseconds). A brief task-irrelevant novel (150 milliseconds) stimulus selected from international affective picture system (Center for the Study of Emotion and Attention, 1999) was presented centrally 350 milliseconds prior to the target. Difference wave reflecting LVF and RVF target processing, with the target ERP from the preceding novel stimuli subtracted (i.e., ERP to novel stimuli not followed by a target is subtracted from ERP to targets preceded by novel stimuli). The target stimuli waveforms with maximal amplitudes are shown for the LVF target from F3 and for the RVF from P3. Significantly enhanced target ERPs were observed in patients with OFCx lesions, with frontal P3 enhancement to LFV targets and posterior N2 enhancement to RVF targets (after Hartikainen et al., 2001).

OFCx seems to exert inhibitory modulatory control over anterior and posterior brain regions. Furthermore this modulation appears to be hemisphere specific. Figure 2 illustrates this hemispheric asymmetry with ERP enhancement subsequent to bilateral orbitofrontal lesion observed, which suggests distinct modulatory effects on the left and the right hemispheres. ERPs to left visual field (LVF) targets showed frontocentral P3 enhancement, while right visual field (RVF) targets were associated with significant parieto-temporal N2 enhancement in orbitofrontal patients. Posterior N2 enhancement may reflect release of posterior association areas from orbitofrontal inhibitory control. Likewise the frontal P3 enhancement may reflect loss of orbitofrontal inhibitory modulation of lateral prefrontal circuitries involved in attentional processes. There are some well-known hemispheric asymmetries in attentional processes such as more frequently observed hemispatial neglect following right hemisphere damage (Mesulam, 1981). Attentional asymmetries in performance due to novel emotional stimuli have been reported in healthy subjects and in patients with OFCx damage (Hartikainen et al., 2000a; Hartikainen et al., 2000b; Hartikainen et al., 2001). Asymmetries in attentional processes and in OFCx-hemisphere interactions may therefore be reflected in these asymmetrically enhanced ERP patterns observed after bilateral orbitofrontal damage.

5.   CONCLUDING REMARKS

ERP lesion studies have provided converging evidence with animal research and brain imaging methods and have helped to clarify the roles of prefrontal cortex in attentional mechanisms. Damage to LPFCx impairs early facilitatory modulation of extrastriate processing of relevant stimuli as well as disrupts inhibitory modulation of distracting irrelevant events. These findings demonstrate the integral role of the LPFCx in voluntary attentional selection and in contributing to reliable neural signal-to-noise ratio in posterior sensory and perceptual brain areas. In addition to early modulation of sensory areas, the LPFCx is involved in voluntary attention, modulating posterior association areas and contributing to P3b when demanding cognitive operations are required. More automated tasks do not seem to rely on LPFCx modulation as evidenced by preserved P3b in a simple detection task after LPFCx damage. In addition to deficits in voluntary attentional mechanisms, involuntary attentional mechanisms that are engaged by novel stimuli are disrupted in lateral prefrontal cortex damage. ERP evidence for significant disruption of novelty processing in lateral prefrontal damage is clear.

In comparison to significant reduction of novelty P3a in lateral prefrontal damage (Knight & Scabini, 1998, Figure 1), the enhanced, rather than diminished, responses to auditory and somatosensory novel stimuli suggest no significant disruption in general novelty processing after orbitofrontal damage (Rule et al., in press). The orbitofrontal cortex seems to modulate, rather than generate, novelty responses. Habituation depends on this modulatory effect, and failure in habituation to auditory and somatosensory novel stimuli is observed in orbitofrontal patients (Rule et al., in press).

The orbitofrontal cortex seems to play a complex and context dependent inhibitory modulatory role that is evident from enhanced late ERP responses to both task-relevant and irrelevant stimuli (Figure 2; Hartikainen et al., 2001; Rule et al., in press). Amplitudes of the late N2 and P3 ERP components to targets that were preceded by novel stimuli were significantly enhanced. Despite bilateral orbitofrontal damage, the pattern of ERP enhancement depended on the field of target presentation. This suggests distinct and lateralized OFCx-hemisphere interactions during attention modulation. Left visual field targets produced enhanced frontal positivity that may reflect release of lateral prefrontal processes, possibly similar to those involved in generating P3a. Right visual field targets produced left posterior N2 enhancement, possibly reflecting resource allocation to target discrimination in the posterior parieto-temporal association areas in the absence of orbital inhibitory modulation of posterior areas. The functional significance of these ERP findings remains to be established. However, it is apparent from these results that there is a clear modulatory effect of orbitofrontal cortex on both anterior and posterior brain structures in attentional processes.

Future fMRI studies on patients with focal brain damage to orbitofrontal cortex may provide a more detailed picture of the specific brain areas involved in these asymmetrically disinhibited ERP responses. Converging results from ERP studies in focal brain damage and fMRI studies on intact brain have consolidated many of the theories related to the role of the prefrontal cortex in attention. To further elucidate the contribution of the lateral and orbital prefrontal cortices in attentional modulation of other brain areas, fMRI studies on patients with frontal damage using similar paradigms to those used in ERP studies may prove useful.

ACKNOWLEDGMENTS

Supported by NINDS grant NS21135. We thank Dr. Keith H. Ogawa, Department of Psychology and John Magaddino Neuroscience Laboratory, Saint Mary's College of California, for his comments on the chapter and Clay Clayworth for figure preparation.

REFERENCES

Akbarian, S., Huntsman, M.M., Kim, J.J., Tafazolli, A., Potkin, S.G., Bunney, W.E., & Jones, E.G. (1995). GABAa receptor subunit gene expression in human prefrontal cortex: comparison of controls and schizophrenics. *Cerebral Cortex, 5*, 550-560.

Akbarian, S., Kim, J.J., Potkin, S.G., Hetrick, W.P., Bunney, W.E., & Jones, E.G. (1996). Maldistribution of interstitial neurons in prefrontal white matter of the brains of schizophrenic patients. *Archives of General Psychiatry, 53*, 425-436.

Alain, C., Woods, D.L., & Knight, R.T. (1998). A distributed cortical network for auditory sensory memory in humans. *Brain Research , 812*, 23-37.

Alho, K., Woods, D.L., Algazi, A., Knight, R.T., & Näätänen, R. (1994). Lesions of frontal cortex diminish the auditory mismatch negativity. *Electroencephalography and Clinical Neurophysiology, 91*, 353-362.

Anderson, S.W., Bechera, A., Damasio, H., Tranel, D., & Damasio, A.R. (1999). Impairments in social and moral behavior related to early damage in human prefrontal corex. *Nature Neuroscience, 2*, 1032-1037.

Bahramali, H., Gordon, E., Lim, C.L., Li, W., Lagapoulus, J., Rennie, C., & Meares, R.A. (1997). Evoked related potentials with and without an orienting reflex. *NeuroReport, 8*, 2665-2669.

Barceló, P., Suwazono, S., & Knight, R.T. (2000). Prefrontal modulation of visual processing in humans. *Nature Neuroscience, 3*, 399-403.

Barch, D.M., Braver, T.S., Sabb, F.W., & Noll, D.C. (2000). Anterior cingulate and the monitoring of response conflict: Evidence from an fMRI study of overt verb generation. *Journal of Cognitive Neuroscience, 12*, 298-309.

Barbas, H. (2000). Connections underlying the synthesis of cognition, memory and emotion in primate prefrontal cortices. *Brain Research Bulletin, 52*, 319-330.

Bartus R.T., & Levere, T.E. (1977). Frontal decortication in Rhesus monkeys: A test of the interference hypothesis. *Brain Research, 119*, 233-248.

Baudena, P., Halgren, E., Heit, G., & Clarke, J.M. (1995). Intracerebral potentials to rare target and distractor auditory and visual stimuli: 3. Frontal cortex. *Electroencephalography and Clinical Neurophysiology, 94*, 251-264.

Bechera, A., Damasio, H., & Damasio, A.R. (2000). Emotion, decision making and the orbitofrontal cortex. *Cerebral Cortex, 10*, 295-307.

Botvinick, M., Nystrom, L.E., Fissell, K, Carter, C.S., & Cohen, J.D. (1999). Conflict monitoring versus selection-for-action in anterior cingulate cortex. *Nature, 402*, 179-181.

Brefczynski, J.A., & DeYoe, E.A. (1999). A physiological correlate of the 'spotlight' of visual attention. *Nature Neuroscience, 2*, 370-374.

Büchel, C., & Friston, K. J. (1997). Modulation of connectivity in visual pathways by attention: Cortical interactions evaluated with structural equation modeling and fMRI. *Cerebral Cortex, 7*, 768-778.

Cavada, C., Company, T., Tejedor, J., Cruz-Rizzolo, R.J., & Reinoso-Suarez, F. (2000). The Anatomical connections of the Macaque monkey orbitofrontal cortex. A review. *Cerebral Cortex, 10*, 220-242.

Center for the Study of Emotion and Attention (CSEA-NIMH) (1999). *The International affective picture system: Digitized photographs.* Gainesville, Florida: Center for Research in Psychophysiology, University of Florida.

Chao, L.L., & Knight, R.T. (1995). Human prefrontal lesions increase distractibility to irrelevant sensory inputs. *NeuroReport, 6*, 1605-1610.

Chao, L.L., & Knight, R.T. (1998). Contribution of human prefrontal cortex to delay performance. *Journal of Cognitive Neuroscience, 10*, 167-177.

Chawla, D., Rees, G., & Friston, K.J. (1999). The physiological basis of attentional modulation in extrastriate visual areas. *Nature Neuroscience, 2,* 671-676.

Clark, V.P., Fannon, S., Lai, S., Benson, R., & Bauer, L. (2000). Responses to rare visual target and distractor stimuli using fMRI. *Journal of Neurophysiology, 83,* 3133-3138.

Corbetta, M. (1998). Frontoparietal cortical networks for directing attention and the eye to visual locations: Identical, independent, or overlapping neural systems? *Proceedings of the National Academy of Sciences USA, 95,* 831-838.

Courchesne, E., Hillyard, S.A., & Galambos, R. (1975). Stimulus novelty, task relevance, and the visual evoked potential in man. *Electroencephalography and Clinical Neurophysiology, 39,* 131-143.

Daffner, K.R., Mesulam, M..M., Holcomb, P.J., Calvo, V., Acar, D., Chabrerie, A., Kikinis, R., Jolesz, F.A., Rentz, D.M., & Scinto L.F. (2000). Disruption of attention to novel events after frontal lobe injury in humans. *Journal of Neurology, Neurosurgery and Psychiatry, 68,* 18-24.

Damasio A.R., (1996). The somatic marker hypothesis and the possible functions of the prefrontal cortex. *Philosophical Transactions of the Royal Society of London. Series B, 351,* 1413-1420.

D'Esposito, M., Detre, J.A., Alsop, D. C., Shin, R. K., Atlas, S., & Grossman, M. (1995). The neural basis of the central executive system of working memory. *Nature, 378,* 279-281.

D'Esposito, M., & Postle, B.R. (1999a). The dependence of span and delayed-response performance on prefrontal cortex. *Neuropsychologia, 37,* 1303-1315.

D'Esposito, M., Postle, B.R., Ballard, D., & Lease, J. (1999b). Maintenance versus manipulation of information held in working memory: An event-related fMRI study. *Brain and Cognition, 41,* 66-86.

D'Esposito, M., Postle, B.R., Jonides, J., & Smith, E.E. (1999c). The neural substrate and temporal dynamics of interference effects in working memory as revealed by event-related functional MRI. *Proceedings of the National Academy of Science USA, 96,* 7514-7519.

Dias, R., Robbins, T.W., & Roberts, A.C. (1996). Dissociation in prefrontal cortex of affective and attentional shifts. *Nature, 380,* 69-72.

Dias, R., Robbins, T.W., & Roberts, A.C. (1997). Dissociable forms of inhibitory control within prefrontal cortex with an analog of the Wisconsin Card Sort Test: Restriction to novel situations and independence from "on-line" processing. *Neuroscience, 17,* 9285-9297.

Dimitrov, M., Phipps, M., Zahn, T.P., & Grafman, J. (1999). A thoroughly modern Gage. *Neurocase, 5,* 345-354.

Donchin, E., & Coles, M.G.H. (1988). Is the P300 component a manifestation of context updating? *Behavioral and Brain Sciences, 11,* 357-427.

Downar, J., Crawley, A.P., Mikulis, D.J., & Davis, K.D. (2000). A multimodal cortical network for the detection of changes in the sensory environment. *Nature Neuroscience, 3,* 277-283.

Dronkers, N.F., Redfern, B.B., & Knight, R. T. (2000). The neural architecture of language disorders. In M. Gazzaniga (Ed.), *The new cognitive neurosciences.* (pp. 949-958). MIT Press.

Escera, C., Alho, K., Winkler, I., & Näätänen, R. (1998). Neural mechanisms of involuntary attention to acoustic novelty and change. *Journal of Cognitive Neuroscience, 10,* 590-604.

Eslinger, P.J. (1999). Orbital frontal cortex: Historical and contemporary views about its behavioral and physiological significance. An introduction to special topic papers. *Neurocase: Case studies in neuropsychology, neuropsychiatry, & behavioural neurology, 5,* 225-229.

Friedman D., Cycowicz Y.M., & Gaeta H. (2001). The Novelty P3: An event-related brain potential (ERP) sign of the brain's evaluation of novelty. *Neuroscience and Behavioral Reviews, 25,* 355-373.

Fuster, J.M., Brodner, M., & Kroger, J.K. (2000). Cross-modal and cross-temporal associations in neurons of frontal cortex. *Nature, 405,* 347-351.

Gehring, W.J., & Knight, R.T. (2000). Prefrontal-cingulate interactions in action monitoring. *Nature Neuroscience, 3,* 516-520.

Gonzalez, C.M.G., Clark, V.P., Fan, S., Luck, S.J., & Hillyard, S.A. (1994). Sources of attention-sensitive visual event-related potentials. *Brain Topography, 7,* 41-51.

Guillery, R.W., Feig, S.L., & Lozsádi, D.A. (1998). Paying attention to the thalamic reticular nucleus. *Trends in Neurosciences, 21,* 28-32.

Halgren, E., Baudena, P, Clarke, J.M., Heit, G., Liégeois, C., Chauvel, P., & Musolino, A. (1995a). Intracerebral potentials to rare target and distractor stimuli: 1.Superior temporal plane and parietal lobe. *Electroencephalography and Clinical Neurophysiology, 94,* 191-220.

Halgren, E., Baudena, P., Clarke, J.M., Heit, G., Marinkovic, K., Devaux, B., Vignal, J.P., & Biraben, A. (1995b). Intracerebral potentials to rare target and distractor stimuli: 2. Medial, lateral and posterior temporal lobe. *Electroencephalalography and Clinical Neurophysiology, 94,* 229-250.

Halgren, E., Marinkovic, K., & Chauvel, P. (1998). Generators of the late cognitive potential in auditory and visual oddball tasks. *Electroencephalography and Clinical Neurophysiology, 106,* 156-164.

Harlow, J.M. (1993). Recovery from the passage of an iron bar through the head. *History of Psychiatry, 4,* 271-281.

Harrington, D.L., Haaland, K.Y., & Knight, R.T. (1998). Cortical networks underlying mechanisms of time perception. *Journal of Neuroscience, 18,* 1085-1095.

Hartikainen, K.M., Ogawa, K.H., & Knight, R.T. (2000a). Transient interference of right hemisphere function due to automatic emotional processing. *Neuropsychologia, 38,* 1576-1580.

Hartikainen, K.M., Ogawa, K.H., Soltani, M., & Knight, R.T. (2000b). Altered emotional influence on visual attention subsequent to orbitofrontal damage in humans [Abstract]. *Society for Neuroscience, 26,* 2023.

Hartikainen, K.M., Ogawa, K.H., Soltani, M., Pepitone, M., & Knight, R.T. (2001). Effects of emotional stimuli on event-related potentials and reaction times in orbitofrontal patients. *Brain and Cognition. 47,* 339-341.

Heinze, H.J., Mangun, G.R., Burchert, W., Hinrichs, H., Scholz, M., Münte, T.F. Gös, A., Scherg, M., Johannes, S., Hundeshagen, H., Gazzanga, M.S., & Hillyard, S.A. (1994). Combined spatial and temporal imaging of brain activity during visual selective attention in humans. *Nature, 372,* 543-546.

Hillyard, S. A., & Anllo-Vento, L. (1998). Event-related brain potentials in the study of visual selective attention. *Proceedings of the National Academy of Sciences USA, 95,* 781-787.

Hornak, J., Rolls, E.T., & Wade, D. (1996). Face and voice expression identification in patients with emotional and behavioural changes following ventral frontal lobe damage. *Neuropsychologia, 34,* 247-261

Hopfinger, J.P., Buonocore, M.H., & Mangun, G.R. (2000). The neural mechanisms of top-down attentional control. *Nature Neuroscience, 3,* 284-291.

Jagust, W. (1999). Neuroimaging and the frontal lobes: Insights from the study of neurodegenerative diseases. In B.L. Miller & J.L. Cummings, (Eds.), *The human frontal lobes: functions and disorders.* (pp. 107-122). New York: Guilford Press.

Jonides, J., Smith, E.E., Koeppe, R.A., Awh, E., Minoshima, S., & Mintun, M.A. (1993). Spatial working memory in humans as revealed by PET. *Nature, 363*, 623-625.

Jonides, J., Smith, E.E., Marshuetz, C., Koeppe, R.A., & Reuter-Lorenz, P.A. (1998). Inhibition of verbal working memory revealed by brain activation. *Proceedings of the National Academy of Sciences USA, 95*, 8410-8413.

Kaipio, M.L., Alho K., Winkler I., Escera C., Surma-aho O., & Näätänen R. (1999). Event-related brain potentials reveal covert distractibility in closed head injuries. *NeuroReport, 10*, 2125-2129.

Katayama J., & Polich J. (1998). Stimulus context determines P3a and P3b. *Psychophysiology, 35*, 23-33.

Kastner, S., Pinsk, M.A., De Weerd, P., Desimone, R., & Ungerleider, L.G. (1999). Increased activity in human visual cortex during directed attention in the absence of visual stimulation. *Neuron, 22*, 751-761.

Kaufer D.I., & Lewis D.A. (1999). Frontal lobe anatomy and cortical connectivity. In B.L. Miller & J.L. Cummings (Eds.), *The human frontal lobes: Functions and disorders.* (pp. 27-44). New York: Guilford Press.

Knight, R.T. (1984). Decreased response to novel stimuli after prefrontal lesions in man. *Electroencephalography and Clinical Neurophysiology, 59*, 9-20.

Knight, R.T. (1996). Contribution of human hippocampal region to novelty detection. *Nature, 383*, 256-259.

Knight, R.T. (1997). Distributed cortical network for visual attention. *Journal of Cognitive Neuroscience, 9*, 75-91.

Knight, R.T., D. Scabini, D.L. Woods, & Clayworth, C.C. (1989). Contribution of the temporal-parietal junction to the auditory P3. *Brain Research, 502*, 109-116.

Knight, R. T., Staines, W. R., Swick, D., & Chao, L.L. (1998). Prefrontal cortex regulates inhibition and excitation in distributed neural networks. *Acta Psychologia, 101*, 159-178.

Knight, R.T., & Nakada, T. (1998). Cortico-limbic circuits and novelty: A review of EEG and blood flow data. *Reviews in Neurosciences, 9*, 57-70.

Knight, R.T., & Scabini, D. (1998). Anatomic bases of event-related potentials and their relationship to novelty detection in humans. *Journal of Clinical Neurophysiology, 15*, 3-13.

Leduc, M., Herron, J.E., Greenberg, D.R., Eslinger, P.J., & Grattan, L.M. (1999). Impaired awareness of social and emotional competencies following orbital frontal lobe damage. *Brain and Cognition, 40*, 174-177.

Levine, B., Freedman, M., Dawson, D., Black, S., & Stuss, D.T. (1999). Ventral frontal contribution to self-regulation: Convergence of episodic memory and inhibition. *Neurocase, 5*, 263-275.

Linden, D.E.J., Prvulovic, D., Formisano, E., Vollinger, M., Zanella, F.E., Goebel, R., & Dierks, T. (1999). The functional neuroanatomy of target detection: An fMRI study of visual and auditory oddball tasks. *Cerebral Cortex, 9*, 815-823.

Macmillan M. (2000). *An odd kind of fame: Stories of Phineas Gage.* Cambridge, MA: MIT Press.

Malmo, R.R. (1942). Interference factors in delayed response in monkeys after removal of frontal lobes. *Journal of Neurophysiology, 5*, 295-308.

Mangun, G.R. (1995). Neural mechanisms of visual selective attention. *Psychophysiology, 32*, 4-18.

Martinez, A., Anllo-Vento L., Sereno, M.I., Frank, L.R., Buxton, R.B., Dubowitz, D.J., Wong E.C., Hinrichs, H., Heinze, H.J., & Hillyard S.A. (1999). Involvement of striate and extrastriate visual cortical areas in spatial attention. *Nature Neuroscience, 2*, 364-369.

McCarthy, G., Luby, M., Gore, J, & Goldman-Rakic, P.S. (1997). Infrequent events transiently activate human prefrontal and parietal cortex as measured by functional MRI. *Journal of Neurophysiology, 77,* 1630-1634.

McDonald, A.W., Cohen, J.D., Stenger, V.A., & Carter, C.S. (2000). Dissociating the role of the dorsolateral prefrontal and anterior cingulate cortex in cognitive control. *Science, 288,* 1835-1838.

McIntosh, A.R., Grady, C.L., Ungerleider, L.G., Haxby, J.V., Rapoport, S.I., & Horwitz, B. (1994). Network analysis of cortical visual pathways mapped with PET. *Journal of Neuroscience, 14,* 655-666.

McIntosh, A.R., Rajah, M.M., & Lobaugh, N.J. (1999). Interaction of prefrontal cortex in relation to awareness in sensory learning. *Science, 284,* 1531-1533.

Menon, K., Ford, J.M., Lim, K.O., Glover, G.H., & Pfefferbaum, A. (1997). A combined event-related fMRI and EEG evidence for temporal-parietal activation during target detection. *NeuroReport, 8,* 3029-3037.

Mesulam, M.M. (1981). A cortical network for directed attention and unilateral neglect. *Annals of Neurology, 10,* 309-325.

Mesulam, M.M. (1986). Frontal cortex and behavior. *Annals of Neurology, 19,* 320-325.

Miller, B.L., Cummings, J.L., Villanueva-Meyer, J., Boone, K., Mehringer, C.M., Lesser, I.M., & Mena, I. (1991). Frontal lobe degeneration: Clinical, neuropsychological, and SPECT characteristics. *Neurology, 41,* 1374-1382.

Müller, N.G., Machado, L., & Knight, R.T. (2002). Contributions of subregions of the prefrontal cortex to working memory: evidence from brain lesions in humans. *Journal of Cognitive Neuroscience, 14,* 673-86.

Nielsen-Bohlman, L., & Knight, R.T. (1999). Prefrontal cortical involvement in visual working memory. *Cognitive Brain Research, 8,* 299-310.

Nies, J.K. (1999). Cognitive and social-emotional changes associated with mesial orbitofrontal damage: Assessment and implications for treatment. *Neurocase, 5,* 313-324.

Näätänen, R. (1992). Attention and brain function. Hillsdale, N.J., Lawrence Erlbaum Associates.

Opitz, B., Mecklinger, A., Friederici, A.D., & von Cramon, D.Y. (1999a). The functional neuroanatomy of novelty processing: Integrating ERP and fMRI results. *Cerebral Cortex, 9,* 379-391.

Opitz, B., Meclinger, A., von Cramon, D.Y., & Krugel, F. (1999b). Combining electrophysiological and hemodynamic measures of the auditory oddball. *Psychophysiology, 36,* 142-147.

Owen, A.M., Stern, C.E., Look, R.B., Tracey, I., Rosen, B.R., & Petrides, M. (1998). Functional organization of spatial and nonspatial working memory processing within the human lateral frontal cortex. *Proceedings of the National Academy of Sciences USA, 95,* 7721-7726.

Paradiso, S., Chemenrinski., E., Yazici., K.M., Tartaro, M., Armando, R., & Robinson, S.G. (1999). Frontal lobe syndrome reassessed: Comparison of patients with lateral or medial frontal brain damage. *Journal of Neurology, Neurosurgery and Psychiatry, 67,* 664-667.

Petersson, K.M., Elfgren, C., & Ingvar, M. (1999). Dynamic changes in the functional anatomy of the human brain during recall of abstract designs. *Neuropsychologia, 37,* 567-587.

Picton T.W. (1992). The P300 wave of the human event-related potential. *Journal of Clinical Neurophysiology, 9,* 456-479.

Polich, J. (1998). P300 clinical utility and control of variability. *Journal of Clinical Neurophysiology, 15,* 14-33.

Prabhakaran, V., Narayanan, K., Zhao, Z., & Gabrieli, J.D. (2000). Integration of diverse information in working memory within the frontal lobe. *Nature Neuroscience, 3,* 85-90.

Price, J.L. (1999). Networks within the orbital and medial prefrontal cortex. *Neurocase, 5,* 231-241.

Raichle M.E., Fiez, J.A., Videen, T.O., MacLeod, A.M., Pardo, J.V., Fox, P.T., & Petersen, S.E. (1994). Practice-related changes in human brain functional anatomy during non-motor learning. *Cerebral Cortex, 4,* 8-26.

Rainer, G., Asaad, W.F., & Miller, E.K. (1998a). Memory fields of neurons in the primate prefrontal cortex. *Proceedings of the National Academy of Sciences USA, 80,* 15008-15013.

Rainer, G., Asaad, W.F., & Miller, E.K. (1998b). Selective representation of relevant information by neurons in the primate prefrontal cortex. *Nature, 393,* 577-579.

Rees, G., Frackowiak, R., & Frith, C. (1997). Two modulatory effects of attention that mediate object categorization in human cortex. *Science, 275,* 835-838.

Roberts, A.C., & Wallis, J.D. (2000). Inhibitory control and affective processing in the prefrontal cortex: neuropsychological studies in the common marmoset. *Cerebral Cortex, 10,* 252-262.

Rolls, E.T. (2000). The orbitofrontal cortex and reward. *Cerebral Cortex, 10,* 284-294.

Rosen, H.J., Hartikainen, K.M., Jagust, W., Kramer, J., Cummings, J., Boone, K., Ellis, W., Miller, C., & Miller, B. (2002). Utility of clinical criteria in differentiating frontotemporal lobar degeneration from AD. *Neurology, 58,* 1608-16015.

Rossi, A.F., Rotter, P.S., Desimone, R., & Ungerleider, L.G. (1999) Prefrontal lesions produce impairments in feature-cued attention. *Society of Neuroscience Abstract, 25,* 3.

Rule, R., Shimamura, A., & Knight, R.T. (in press). Orbitofrontal cortex and dynamic filtering of emotions. *Cognitive, Affective and Behavioral Neuroscience.*

Rypma, B., & D'Esposito, M. (2000). Isolating the neural mechanisms of age-related changes in human working memory. *Nature Neuroscience, 3,* 509-515.

Stam, C.J., Visser, S.L., Op de Coul, A.A., De Sonneville, L.M., Schellens, R.L., Brunia, C.H., de Smet, J.S., & Gielen, G. (1993). Disturbed frontal regulation of attention in Parkinson's disease. *Brain, 116,* 1139-1158.

Stern, C.E., Corkin S., Gonzalez, R.G., Guimares, A.R., Baker, J.R., Jennings, P.J., Carr, C.A., Sugiura, R.M., Vedantham, V., & Rosene, B.R. (1996). The hippocampal formation participates in novel picture encoding: Evidence from functional magnetic resonance imaging. *Proceedings of the National Academy of Sciences USA, 93,* 8660-8665.

Stone, V.E., Baron-Cohen, S., & Knight, R.T. (1998). Does frontal lobe damage produce theory of mind impairment? *Journal of Cognitive Neuroscience, 10,* 640-656.

Stuss, D.T., Levine, B., Alexander, M.P., Hong, J., Palumbo, C., Hamer, L., Murphy, K.J., & Izukawa, D. (2000). Wisconsin Card Sorting Test performance in patients with focal frontal and posterior brain damage: effects of lesion location and test structure on separable cognitive processes. *Neuropsychologia, 38,* 388-402.

Swick, D. (1998). Effects of prefrontal lesions on lexical processing and repetition priming: An ERP study. *Cognitive Brain Research, 7,* 143-157.

Swick, D., & Knight, R.T. (1999). Contributions of prefrontal cortex to recognition memory: Electrophysiological and behavioral evidence. *Neuropsychology, 13,* 155-170.

Tulving, E., Markowitsch, H.J., Kapur, S., Habib, R., & Houle, S. (1994). Novelty encoding networks in the human brain: Positron emission tomography data. *NeuroReport, 5,* 2525-2528.

Tulving, E., Markowitsch, H.J., Craik, F.I.M., Habib, R., & Houle, S. (1996). Novelty and familiarity activations in PET studies of memory encoding and retrieval. *Cerebral Cortex, 6,* 71-79.

Verleger, R., Heide, W., Butt, C., & Kompf, D. (1994). Reduction of P3b potentials in patients with temporo-parietal lesions. *Cognitive Brain Research, 2,* 103-116.

Weinberger, D.R., Berman, K.F., & Zec, R.F. (1986) Physiological dysfunction of dorsolateral prefrontal cortex in schizophrenia. *Archives of General Psychiatry, 43,* 114-124.

Weinberger, D.R., Berman, K.F., Suddath, R., & Torrey, E.F. (1992). Evidence of dysfunction of a prefrontal-limbic network in schizophrenia: A magnetic resonance imaging and regional cerebral blood flow study of discordant monozygotic twins. *American Journal of Psychiatry, 149,* 890-897.

Wilkins, A.J., Shallice, T., & McCarthy, G. (1987). Frontal lesions and sustained attention. *Neuropsychologia, 25,* 359-365.

Woods, D.L., & Knight, R.T. (1986). Electrophysiological evidence of increased distractibility after dorsolateral prefrontal lesions. *Neurology, 36,* 212-216.

Yamaguchi, S., & Knight, R.T. (1991). Anterior and posterior association cortex contributions to the somatosensory P300. *Journal of Neuroscience, 11,* 2039-2054.

Yamaguchi, S., & Knight, R.T. (1992). Effects of temporal-parietal lesions on the somatosensory P3 to lower limb stimulation. *Electroencephalography and Clinical Neurophysiology, 84,* 139-148.

Chapter 7

ERP AND fMRI CORRELATES OF TARGET AND NOVELTY PROCESSING

BERTRAM OPITZ
Experimental Neuropsychology Unit, Saarland University, 66041 Saarbrücken, Germany

1. INTRODUCTION

Human behavior presupposes the ability to respond actively to biologically significant events. One important behavior that occurs within the context of the interaction of an organism and the environment is the detection of changes. Such change detection may be regarded as the outcome of processes that include extraction of stimulus information, reallocation of attention, and sensory memory. Discrete lesions in frontal, temporo-parietal, or medio-temporal cortices, can disrupt the behavioral processes associated with different aspects of change detection. What lesion studies cannot inform, however, is which regions in the normal brain subserve the detection of a change. A combined analysis of two major methods in cognitive neuroscience, event-related potentials (ERPs) and functional magnetic resonance imaging (fMRI), can address the issue of spatio-temporal characteristics of normal brain activation underlying change detection.

2. TWO PROCESSES, MULTIPLE BRAIN STRUCTURES

The present chapter considers change detection processes that are triggered by events that are distinct from the majority of the occurring events and either signal task relevant information (target detection) or a perceptually novel event (novelty detection). Change detection processes are often assessed by means of the so-called "oddball task". This procedure requires subjects to attend to a stimulus stream and to discriminate between frequently occurring regular events and rare, irregular events (Donchin et al., 1997). The regular events are called standards and the irregular events are referred to as targets, because in most studies a covert or overt response (mental counting, a button press) is required to induce task relevance of the stimulus events. When such a target is consciously detected, a parietally distributed positive ERP component is generated that is known as the P300 or P3b (Sutton et al., 1965; Donchin, 1981; Donchin & Coles, 1988). There is accumulating evidence from neurophysiological studies (Knight, 1996; Halgren et al., 1995) that several brain structures are involved in different aspects of target detection. These structures include the medial temporal lobe, the frontal cortex, the supramarginal gyrus, and the anterior cingulate gyrus. Studies using fMRI also have found activation of the lateral prefrontal cortex, the insular cortex bilaterally, and subcortical structures, such as the thalamus during target detection processes in the auditory and visual modalities (McCarthy et al., 1997; Linden et al., 1999). In addition, source localization performed on the basis of magnetoecephalographic data suggests contributions from subcortical structures (Mecklinger et al., 1998).

As with target detection, novelty detection processes can be examined in the visual (Courchesne et al., 1975), auditory, and somatosensory (Knight, 1996; Polich et al., 1991) modalities. Novel events cause an involuntary attentional shift because of their potential behavioral significance, even when they occur outside the current focus of attention (Knight, 1984). The neurophysiological mechanisms underlying the detection of novelty are reflected in the novelty P3 ERP. Compared to the P3b elicited by targets, the novelty P3 has a shorter latency and a more frontal scalp topography distribution (cf. Friedman & Simpson, 1994).

The novelty P3 scalp topography appears to reflect the activity of a widespread neuronal network including the frontal and the parietal lobes, as well as lateral and medial temporal lobe structures (Alho et al., 1998; Mecklinger & Ullsperger, 1995). Additionally, research on epileptic patients using depth electrode measurements also suggests that the novelty P3 reflects the activity of a distributed network, with major components in the hippocampus, the temporal lobes and dorsolateral prefrontal cortex (Baudena et al., 1995; Halgren et al., 1995).

Taken together, these findings indicate that the processes involved in target and novelty detection constitute a widespread and partially overlapping neuronal network. However, the exact neuronal substrates of these change detector networks are not completely understood. Until recently, electrophysiological methods such as ERPs and hemodynamic approaches like fMRI were used separately to disentangle the neuroanatomical structures mediating change detection. The present chapter reviews findings from the use of these methods together.

3. AN INTEGRATED APPROACH TO THE DETECTION OF CHANGE

Change detection processes occur within seconds and, therefore, require an online measure with millisecond time resolution, such as that provided by ERPs. These neuroelectric measures are extracted from the ongoing electroencephalogram (EEG) by signal averaging and are small voltage changes that are time-locked to sensory, motor, or cognitive events (Rugg & Coles, 1995). The resulting waveform can be described as a series of positive or negative deflections called ERP components with a specific latency and amplitude distribution or topography over the scalp. These components index the timing and sequence of neuronal activity elicited by a particular event. A class of ERP components can be elicited by the detection of an event deviating in some manner from the other stimulus events presented in the experiment (Donchin et al., 1997). However, these change detection components differ among themselves in the nature of the deviating event (e.g., small frequency deviants or perceptually novel events) or in the extent to which they are elicited by attended (e.g., the P3b) and/or unattended (novelty P3) events (Friedman et al., 1998).

Although the temporal resolution of the ERP is excellent, its capability in identifying the generating neural sources of the functionally relevant brain structures is necessarily approximate. The difficulty is to solve the "inverse problem," which determines the three dimensional distribution of active areas inside the brain, based solely on two dimensional external electric or magnetic measurements. Given a finite number of electrodes at which the potentials are measured, there exist an infinite number of possible solutions of the inverse problem and, therefore, a potentially infinite number of brain structures can account for the generation of the measured scalp potentials (Koles, 1998). Despite this difficulty, source analyses of the P3b and the novelty P3 have been performed (cf. Alho et al., 1998; Mecklinger & Ullsperger, 1996; Mecklinger et al., 1998). Indeed, if the problem is approached by utilizing prior information, one can substantially reduce the number of possible solutions. For example, anatomical knowledge can be used to constrain sources to locations around the cortical fold (Scherg & Berg, 1991). In addition, the activity of neighboring neurons is more likely synchronized compared to the activity of neurons that are far from each other. In mathematical terms, the task is to find the smoothest of all possible solutions of a distributed activity throughout the brain. This method is called low resolution electromagnetic tomography (LORETA, Pascal-Marqui et al., 1994) and has been employed successfully (Anderer et al., 1998).

An alternative approach to the neuronal localization of cognitive processes is provided by other neuroimaging methods such as fMRI. The measurement technique is based on changes in local blood oxygenation and flow that occur with changes in neural activity. When the appropriate imaging parameters are chosen, fMRI is able to detect changes in blood oxygenation level since oxygenated hemoglobin has much smaller magnetic susceptibility than deoxygenated hemoglobin. This outcome is known as Blood Oxygenation Level Dependent Effect or BOLD signal (Ogawa & Lee, 1990). As suggested by anatomical and physiological evidence, fMRI measures are limited by the spatial extent of the circulatory system detected at a given field strength. Hence, the higher the field strength, the higher the spatial resolution, with spatial resolution as high as 1 mm possible (Cohen, 1996). The temporal resolution is limited by the properties of the brain's circulatory system. Most changes of this hemodynamic response that are detectable with fMRI appear after a delay of several seconds and take about 6 seconds to reach maximum. Thus, despite the excellent spatial resolution, fMRI does not provide the temporal resolution required to make inferences about the subprocesses involved in change detection (Menon & Kim, 1999).

In an effort to overcome the intrinsic limitations of each approach, both the electrophysiological and hemodynamic measures of change detection processes were integrated using neuroanatomically constrained source analysis (cf. Opitz et al., 1999a). The fundamental assumption underlying this method was that the same brain areas that generate the P3b or the novelty P3 in the ERP also would show an increased fMRI BOLD response. The fMRI results can then be used to constrain the inverse problem by providing the number and locations of possible sources for the associated scalp potentials (Mecklinger, 2000). Thus, by modeling the neuronal electric activity employing these neuroanatomical constraints, brain structures can be identified that are involved in change detection while also monitoring the temporal dynamics of the same neural locations.

For a combined analysis of fMRI and ERP, it would be useful to simultaneously record both types of data. A problem with this approach is that the changing magnetic field produced by the MR-scanner, in addition to the electrical wire moving due to heart beat-related body movements, induces electric currents that affect the EEG signal amplifier. It is therefore technically difficult to record artifact-free ERPs within the magnetic field of the scanner (Allen et al., 1998). For this reason, parallel data acquisition in separate sessions with the same subjects employing identical experimental procedures can be used. However, with the improvement of recording equipment and new signal processing methods the scope of combined ERP/fMRI studies can be extended considerably (Allen et al., 2000).

Although the fMRI constrained source analysis has a physiological basis, it is not reasonable to assume that the distribution of fMRI and ERP signals will always perfectly match. To exclude all but the one correct solution, it is important that a derived source model is subjected to statistical evaluation and fulfills a number of criteria. First, it should explain the empirical electrophysiological data very well by accounting for as much of the experimental variance as possible (typically greater than 90%). Second, the estimated source configuration should be in agreement with structural and physiological restrictions. That is, the ERP data are assumed to arise from electrical activity in the gray matter with a dipolar orientation perpendicular to the cortical surface. Third, a limited number of neurons cannot produce an unlimited electric field outside the head. According to realistic estimates based on measured current densities, a dipole strength of 10 nAm would correspond to 40 mm^2 of active cortex (Freeman, 1975). As a consequence, a brain area of a certain spatial extent, for example 30200 mm^2, while showing an increased BOLD response, nonetheless has to be excluded as an ERP generator if the activation strength of the respective dipole exceeds a specific threshold, or 50 nAm in this example. One reason for the absence of an ERP signal even though a BOLD signal is present, is that scalp recorded ERPs are generated only by a synchronous modulation of a neuronal population in an "open field" configuration (Nunez, 1981), whereas a hemodynamic response can be caused by any configuration of neuronal activity. Conversely, neural activity as reflected in the ERPs might not have a detectable hemodynamic counterpart. For instance, this may be the case when relatively few neurons are synchronously highly active but the regional vascular bed is sparse, as is the situation in the hippocampus. The experimental variance unexplained by a generator model constrains the likelihood estimate of such activity: the more unexplained variance, the more likely that sources are missing. Therefore, this combined analysis is a probabilistic approach (Opitz et al., 1999b; Luck, 1999).

Finally, the model has to be specific for a particular ERP component, such that it explains the experimental variance of this component but not of others. Moreover, the activation strength of the source model should also be maximal in the time range of the specific ERP, thereby indicating a maximal contribution solely to this component. This specificity of a source model can be estimated by calculating the time course of dipole strength and goodness-of-fit over the entire period of measurement.

4.   AN EXAMPLE OF COMBINED ANALYSIS

The application of this integrated approach is illustrated in two experiments that employed either task-relevant targets or perceptually novel auditory stimuli. The electrophysiological and hemodynamic brain responses were measured in healthy volunteers in two separate sessions of identical task situations: a stimulus train comprised of different auditory stimuli where the rare targets had to be silently counted by the subjects (Opitz et al., 1999a). In the second experiment, novel stimuli were included in the stimulus train (Opitz et al., 1999b). Findings from studies investigating the functional similarity in the processing of environmental novel sounds and words suggest that environmental sources, like words, can activate conceptual-semantic representations (van Petten & Rheinfelder, 1995). Thus, the novel stimuli were divided into two groups: identifiable novel sounds, which were reliably identified by subjects (e.g., telephone bell or dog bark) and nonidentifiable novels, which were not (Mecklinger et al., 1997). In order to assess temporal aspects of the processing network identified in fMRI activation maps, a neuroanatomically constrained dipole analysis was employed. Dipole locations were kept fixed according to the fMRI activation foci averaged across subjects, whereas dipole orientations were allowed to vary to model the ERP data (Opitz et al., 1999b).

Figure 1 shows the ERPs for midline electrodes in response to the targets, the identifiable novels, and the nonidentifiable novels. As expected, targets elicited a parietally maximal P3b component, with a peak latency of 360 ms at the Pz recording site. Conversely, both types of novels elicited fronto-centrally focused novelty P3s peaking around 280 ms. Notably, identifiable compared to nonidentifiable novel sounds elicited a parietal negative-going deflection at right parietal recording sites peaking around 420 ms. In light of the temporal and topographical similarities with the N400 component produced by semantically unexpected language stimuli this negativity can be described as an N400-like component (Kutas & Hillyard, 1983).

Figure 2 illustrates the overall fMRI data. Four significant clusters of BOLD increase were obtained with peaks in the posterior part of the left and right superior temporal gyri (STG), adjacent to the supramarginal gyri (SMG) and the left and right neostriatum. For modeling the ERP waveforms, generators at all four locations derived from the fMRI activation pattern were first assumed. However, dipoles located at bilateral neostriatum showed an activation strength of more than 200 nAm. Based on the measurements by Freeman (1975), this activation strength would correspond to 8cm^2 of active cortex. Since the neostriatum is much smaller, the respective dipoles were excluded from the final dipole solution.

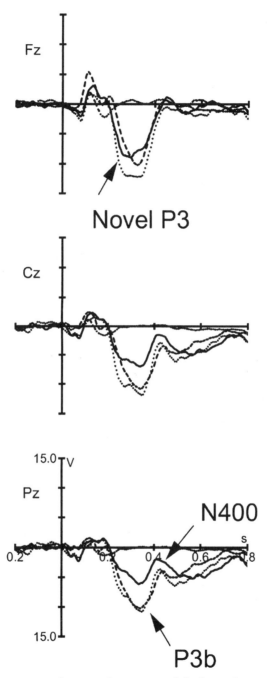

Figure 1. Grand average waveforms to the targets and both novel types at the midline electrode. Solid line=identifiable novels, dotted line=nonidentifiable novels, dashed line=deviant tones. Negative number on top, positive number on bottom.

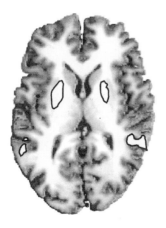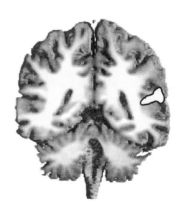

Figure 2. Brain areas that showed significant fMRI activation to target stimuli were superimposed on an individual brain in Talairach space: left and right superior temporal gyri (STG), left supramarginal gyrus (SMG) and neostriatum. Horizontal (left) and coronal sections (right) were thresholded at Z=3.01 (p < .001, one tailed) (after Opitz et al., 1999a).

Figure 3 presents the goodness-of-fit and the activation waveforms of the derived solution. This source analysis suggests that only the bilateral activation of the posterior part of the superior temporal gyrus contributes to the scalp P3b elicited by auditory stimuli. This finding is consistent with previous findings (Menon et al., 1997) and source localization performed on the basis of MEG data (Alho et al., 1998), which suggest contributions from posterior temporal cortex.

In the second experiment, the fMRI activation pattern associated with auditory novelty processing was comprised of two significant clusters of activity in the midportion of the left and right STG. Figure 4 illustrates the fMRI findings. No significant differences in mean fMRI activation size and or specific lateralization could be obtained for either novel type. Furthermore, this STC activity was located anterior to the hemodynamic response to target tones. This outcome could account for the distinctive scalp distribution of the novelty P3. Consistent with this result, a contribution of anterior temporal cortex activity to the novelty P3 has been observed in an MEG study (Alho et al., 1998). Moreover, extensive temporo-parietal lesions centered in the superior temporal cortex attenuated the P3 to novel sounds, especially at posterior recordings (Knight et al., 1989). Despite this converging evidence for superior temporal gyrus contributions to novelty processing, there is a lack of consistency with respect to hippocampal involvement in these processes. Human lesion studies (Knight, 1996) and neuroimaging studies (Martin, 1999; Tulving et al., 1996) have suggested an important role of the hippocampal/parahippocampal region in the processing of novel information. The present data, in agreement with previous P3b studies did not reveal significant activation within or in the vicinity of this brain area (Polich & Squire, 1993). At a functional level a possible explanation of the discrepancy between the present and former data could be derived from neuroimaging studies that demonstrated the involvement of hippocampal structures in a large variety of cognitive processes, such as memory encoding and retrieval (Desgranges et al., 1998). It is conceivable that such processes were also produced by target and standard tones and could therefore have masked the effects of novelty detection in the hippocampus.

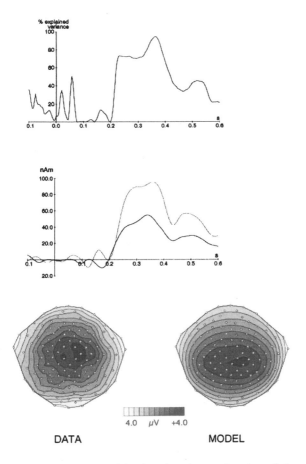

DATA   MODEL

Figure 3. Goodness of fit (% explained variance) as a function of time (top panel), time course of dipole strength (nAm) (middle panel). The scalp potential maps of the fit interval for empirical data (left) and the dipole model (right) are shown. Solid-line=left superior temporal gyrus, dotted line=right superior temporal gyrus (after Opitz et al., 1999a).

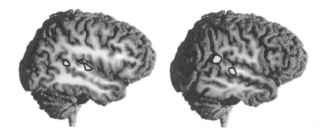

Figure 4. Comparison of fMRI activation to novel sounds (left) and target tones (right) in sagittal section through the right temporal lobe. The activations differ along the anterior-posterior axis of the superior temporal gyrus, with novel sounds activity located anteriorly.

Given the differential ERP waveforms obtained for identifiable and nonidentifiable novels, the absence of any differences in the fMRI activation pattern was surprising. A median split of the sample was constructed according to the most salient ERP differences between identifiable and nonidentifiable novels, and the N400-effect was only obtained for identifiable novels. Subjects showing a large N400-effect will be referred to as the large difference group (LD-group), while the other subjects, with a weak or no N400-effect formed the small difference group (SD-group). Figure 5 displays the ERP waveforms of both novel stimuli at a representative electrode from the right parietal region and the scalp topography of the N400-effect for both groups. When the fMRI data analysis was conducted separately for the LD and SD groups, the activation pattern for identifiable and nonidentifiable novels was clearly dissociable for the LD-group, but not for the SD-group. Figure 6 illustrates those effects. In the LD-group bilateral activation of the midportion of the STC was obtained for the nonidentifiable and identifiable novels. In this group, additional right frontal activation for identifiable novels was found.

Based on these results it can be hypothesized that (a) bilateral activation of the middle STG accounts for the generation of the auditory novelty P3 whereas (b) an additional right frontal generator might contribute solely to the ERP waveforms for identifiable novels. To test these hypotheses, the scalp ERP distribution of the LD-group was modeled, for which reliable ERP differences between both novel types were obtained, using dipole source locations in both STG and in the right frontal area, as derived from functional images.

Examination of the time course of dipole activation in the LD group revealed that the dipoles in the STG for identifiable as well as nonidentifiable novels explained more than 90% of the signal variances within the time range from 260 to 360 ms only. Figure 7 summarizes the findings. These results support the view that the middle STG is one of the contributors to the scalp recorded novelty P3. The third dipole in this model accounted for 47% of the variance in the ERP elicited by identifiable novels but only 15% in the ERP elicited by nonidentifiable novels. The right frontal dipole showed a clear maximum of activation strength subsequent to the novelty P3 for identifiable but not for nonidentifiable novels. This outcome is consistent with the ERP data and further strengthens the observation that the right frontal BOLD response is apparent for identifiable but negligible for nonidentifiable novels.

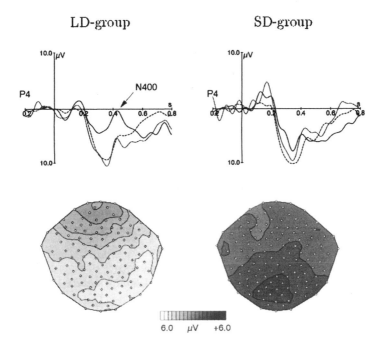

Figure 5. ERP waveforms for a representative electrode of the right parietal region (upper panel) and the scalp distribution (lower panel) of the N4-effect for the LD (left) and SD groups (right) are shown. Dark areas in the potential maps indicate positive differences and light areas negative differences between identifiable and nonidentifiable novel sounds. Negative number on top, positive number on bottom.

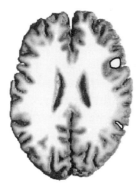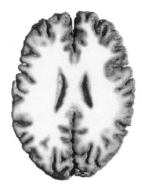

Figure 6. BOLD changes in the prefrontal cortex in response to identifiable (left) and nonidentifiable (right) novel stimuli for the LD-group obtained by averaging the groups separately. Note that the slice where right frontal activation was obtained for identifiable novels is also shown for the nonidentifiable novels although no activation was present (after Opitz et al., 1999b).

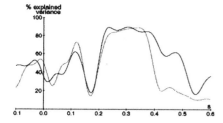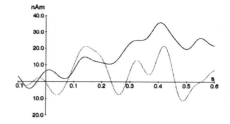

Figure 7. Percent explained variance (left) and dipole strength (right) of the dipole solution of the right frontal dipole for identifiable (solid tracer) and nonidentifiable (dotted trace) novels (after Opitz et al., 1999b).

The involvement of prefrontal cortex in novelty processing was suggested by neuropsychological studies showing a decreased ERP response to auditory and visual novel stimuli after unilateral prefrontal lesions (Knight, 1984). However, the lack of selective amplitude reduction of the novelty P3 over the lesioned hemisphere led Knight (1984) to conclude that the prefrontal cortex is not the primary generator of these brain potentials, and that this activity more likely reflects attentional modulations of generators located elsewhere (e.g., in the superior temporal cortex). More evidence for the notion that the prefrontal cortex might act as an attentional control system was provided by a recent study that compared the novelty P3 in young and old adults (Friedman et al., 1998). They found a reduced novelty P3 for old as compared to young adults, in a condition where subjects had to attend to the stimulation but had showed identical novelty P3s for young and old adults in an ignore condition. This finding suggests that prefrontal activity during an attended novelty oddball task more likely reflects attentional modulations rather than a true generator of the scalp recorded novelty P3. Furthermore, as none of these studies reported a specific lateralization of frontal lobe involvement in novelty processing, processes other than novelty detection seem to be reflected in lateralized PFC activity in general, and in the rPFC found in the present study, in particular. The results of the combined analyses suggest that the rPFC is activated only for those novel events that activate a semantic concept and that this activation is delayed relative to processes underlying novelty detection. In recent functional imaging studies, rPFC activation of a similar kind has been found when previously learned material had to be retrieved from episodic memory (Brewer et al., 1998; Opitz et al., 2000). Moreover, it has been demonstrated that the activity of the prefrontal cortex in such memory tasks is modulated by intrinsic stimulus properties (i.e., identifiable and nonidentifiable sounds) as well as task demands (Opitz et al., 2000). In light of these findings, the activity of the rPFC in the present study appears to reflect the encoding and/or retrieval of conceptual semantic information carried by identifiable novel sounds.

5. CONCLUSION

The data presented together with those of previous studies point to differential brain networks underlying target and novelty detection. The combined analyses of ERP and fMRI data suggest a major contribution of the superior temporal gyrus to the generation of the P3b and the novelty P3 with the latter being generated more anterior loci. Moreover, this combined analysis supports the view that novelty processing consists of at least two sequential subprocesses: first, an automatically operating novelty detection mechanism, subserved by superior temporal structures and second, further processes based on a novel sounds' meaning, subserved by right frontal cortical areas. The precise nature of the psychological process reflected in this rPFC activation remains to be elucidated. Nevertheless, the present study demonstrates that this integrated approach provides a new opportunity to disentangle the temporal aspects of neural activation underlying auditory target and novelty detection. The remaining question is whether the same neuronal network underlying auditory change detection will also subserve change detection in the visual domain. A recent neuroimaging study has shown that similar brain areas including the supramarginal gyrus, insular cortex bilaterally, and circumscribed parietal regions are activated during auditory and visual target detection (Linden et al., 1999). However, it remains unclear which of these brain structures contribute to the scalp recorded potentials in the respective stimulus modality. Future dipole analyses with neuroanatomical constraints can be utilized to tackle this problem.

ACKNOWLEDGMENTS

Figure 6 reprinted from Opitz, B., Mecklinger, A., Friederici, A.D., & von Cramon, D.Y. (1999b). The functional neuroanatomy of novelty processing: Integrating ERP and fMRI results. *Cerebral Cortex, 9,* 379-391. Copyright (2002) with permission from Oxford University Press.

REFERENCES

Alho, K., Winkler, I., Escera, C., Huotilainen, M., Virtanen, J., Jaaskelainen, 1. P., Pekkonen, E., & Ilmoniemi, R.J. (1998). Processing of novel sounds and frequency changes in the human auditory cortex: Magnetoencephalographic recordings. *Psychophysiology, 35,* 211-224.

Allen, P.J., Josephs, 0. & Turner, R. (2000). A method for removing imaging artifact from continuous EEG recorded during functional MRI. *Neuroimage, 12,* 230-239.

Allen, P.J., Polizzi. C., Krakow, K., Fish. D.R., & Lemieux, L. (1998). Identification of EEG events in the MR scanner: The problem of pulse artifact and a method for its subtraction. *Neuroimage, 8,* 229-239.

Anderer, P., Pascual-Marqui, R.D., Semlitsch, H.V., & Saletu, B. (1998). Differential effects of normal aging on sources of standard N1, target Nl and target P300 auditory event related brain potentials revealed by low resolution electromagnetic tomography (LORETA). *Electroencephalography and Clinical Neurophysiology, 108,* 160-174.

Baudena, P., Halgren, E., Heit, G., & Clarke, J.M. (1995). Intracerebral potentials to rare target and distractor auditory and visual stimuli. III: Frontal cortex. *Electroencephalography and Clinical Neurophysiology, 94,* 251-264.

Brewer, J.B., Zhao, Z., Desmond, J.E., Glover, C.H., & Gabrieli, J.D.E. (1998). Making memories: Brain activity that predicts how well visual experience will be remembered. *Science, 281,* 1185-1187.

Cohen, M.S. (1996). Rapid MRI and functional applications. In A. W. Toga & J. C. Mazziotta (Eds.), *Brain mapping: The methods* (pp. 223-255). San Diego: Academic Press.

Courchesne, E., Hillyard, S.A., & Galambos, R. (1975). Stimulus novelty, task relevance and the visual evoked potential in man. *Electroencephalography and Clinical Neurophysiology, 39,* 131-143.

Desgranges, B., Baron, J.-C., & Eustache, F. (1998). The functional neuroanatomy of episodic memory: The role of the frontal lobes, the hippocampal formation, and other areas. *Neuroimage, 8,* 198-213.

Donchin, E. (1981). Surprise! ... Surprise? *Psychophysiology, 18,* 493-515.

Donchin, E., & Coles, M.G.H. (1988). Is the P300 component a manifestation of context updating? *Behavioral Brain Science, 11,* 357-372.

Donchin, E., Spencer, K., & Dien, J. (1997). The varieties of deviant experience: ERP manifestations of deviance processors. In G.J.M. van Box & K.B.E. Becker (Eds.), *Brain and behavior: Past, present and future* (pp. 67-91). Tilburg: Tilburg University Press.

Freeman, W. J. (1975). *Action in the nervous system.* New York: Academic Press.

Friedman, D., Kazmerski, V.A., & Cycowicz, Y.M. (1998). Effects of aging on the novelty P3 during attend and ignore oddball tasks. *Psychophysiology, 35,* 508-520.

Friedman, D., & Simpson, G.V. (1994). ERP amplitude and scalp distribution to target and novel events: effects of temporal order in young, middle-aged and older adults. *Cognitive Brain Research, 2,* 49-63.

Halgren, E., Baudena, P., Clarke, J.M., Heit, G., Liegeois, C., & Musolino, A. (1995). Intracerebral potentials to rare target and distractor auditory and visual stimuli. 1: Superior temporal and parietal lobe. *Electroencephalography and Clinical Neurophysiology, 94,* 191-220.

Knight, R.T. (1984). Decreased response to novel stimuli after prefrontal lesions in man. *Electroencephalography and Clinical Neurophysiology, 59,* 9-20.

Knight, R.T. (1996). Contribution of human hippocampal region to novelty detection. *Nature, 383,* 256-259.

Knight. R.T., Scabini, D., Woods, D.L., & Clayworth, C. (1989). Contributions of the temporal parietal junction to the human auditory P3. *Brain Research, 502,* 109-116.

Koles, Z.J. (1998). Trends in EEG source localization. *Electroencephalography and Clinical Neurophysiology, 106,* 127-137.

Kutas, M., & Hillyard, S.A. (1983). Event-related brain potentials to grammatical errors and semantic anomalies. *Memory and Cognition, 11,* 539-550.

Linden, D.E.J., Prvulovic, D., Formisano, E., Vollinger, M., Zanella, F.E., Goebel, R., & Dierks, T. (1999). The functional neuroanatomy of target detection: An fMRI study of visual and auditory oddball tasks. *Cerebral Cortex, 9,* 815-823.

Luck, S.J. (1999). Direct and indirect integration of event-related potentials, functional magnetic resonance images and single-unit recordings. *Human Brain Mapping, 8,* 115-120.

Martin. A. (1999). Automatic activation of the medial temporal lobe during encoding: Lateralized influences of meaning and novelty. *Hippocampus, 9,* 62-70.

McCarthy, G., Luby, M., Gore, J., & Goldman-Rakic, P. (1997). Infrequent events transiently activate human prefrontal and parietal cortex as measured by functional MRI. *Journal of Neurophysiology, 77,* 1630-1634.

Mecklinger, A. (2000). Interfacing mind and brain: A neurocognitive model of recognition memory. *Psychophysiology, 37,* 565-582.

Mecklinger, A., Maess, B., Opitz, B., Pfeifer, E., Cheyne, D., & Weinberg, H. (1998). A MEG analysis of the P300 in visual discrimination task. *Electroencephalography and Clinical Neurophysiology, 108,* 45-56.

Mecklinger, A., Opitz, B., & Friederici, A.D. (1997). Semantic aspects of novelty detection in humans. *Neuroscience Letters, 235,* 65-68.

Mecklinger, A., & Ullsperger, P. (1995). The P300 to novel and target events: a spatio-temporal dipole model analysis. *NeuroReport, 7,* 241-245.

Menon, R.S., & Kim, S. (1999). Spatial and temporal limits in cognitive neuroimaging with fMRI. *Trends in Cognitive Sciences, 3,* 207-216.

Menon, V., Ford, J.M., Lim, K.O., Glover, G.H., & Pfefferbaum, A. (1997). Combined event-related fMRI and EEG evidence for temporal-parietal cortex activation during target detection. *NeuroReport, 8,* 3029-3037.

Nunez, P.L. (1981). *Electric fields of the brain: The neurophysics of EEG*. New York: Oxford University Press.

Ogawa. S., & Lee, T.A (1990). Magnetic resonance imaging of blood vessels at high fields in-vivo and in-vitro measurements and image simulation. *Magnetic Resonance in Medicine, 16*, 9-18.

Opitz, B., Mecklinger, A., von Cramon, D.Y., & Kruggel, F. (1999a). Combining electrophysiological and hemodynamic measures of the auditory oddball. *Psychophysiology, Special Report, 36*, 142-147.

Opitz, B., Mecklinger, A., & Friederici, A.D. (2000). Functional asymmetry of human prefrontal cortex: Encoding and retrieval of verbally and nonverbally coded information. *Learning and Memory, 7*, 85-96.

Opitz, B., Mecklinger, A., Friederici, A.D., & von Cramon, D.Y. (1999b). The functional neuroanatomy of novelty processing: Integrating ERP and fMRI results. *Cerebral Cortex, 9*, 379-391.

Pascal-Marqui, R.D., Michel, C.M., & Lehmann, D. (1994). Low resolution electromagnetic tomography: A new method for localizing electrical activity in the brain. *International Journal of Psychophysiology, 18*, 49-65.

van Petten, C., & Rheinfelder, H. (1995). Conceptual relationships between spoken words and environmental sounds: Event-related brain potentials measures. *Neuropsychologia, 33*, 485-508.

Polich, J., Brock, T., & Geisler, M. (1991). P300 from auditory and somatosensory stimuli: Probability and inter-stimulus interval. *International Journal of Psychophysiology, 11*, 219-223.

Polich, J., & Squire, L.R. (1993). P300 from amnestic patients with bilateral hippocampal lesions. *Electroencephalography and Clinical Neurophysiology, 86*, 408-417.

Rugg, A.D., & Coles, M.G.H. (1995). *Electrophysiology of mind: Event-related brain potentials and cognition*. New York: Oxford University Press.

Scherg, M., & Berg, P. (1991). Use of prior knowledge in brain electromagnetic source analysis. *Brain Topography, 4*, 143-150.

Sutton, S., Braren, M., Zubin, J., & John. E.R. (1965). Evoked-potential correlates of stimulus uncertainty. *Science, 150*, 1187-1188.

Tulving, E., Markowitsch, H.J., Craik, F.I.M., Habib, R., & Houle, S. (1996). Novelty and familiarity activations in PET studies of memory encoding and retrieval. *Cerebral Cortex, 6*, 71-79.

Chapter 8

EEG AND ERP IMAGING OF BRAIN FUNCTION

ALAN GEVINS, MICHAEL E. SMITH, AND LINDA K. McEVOY
San Francisco Brain Research Institute and SAM Technology

1. INTRODUCTION

The purpose of functional brain mapping is to localize patterns of neuronal activity associated with sensory, motor, and cognitive functions, or with disease processes. To be complete, an imaging modality needs near millimeter precision in localizing regions of activated tissue and sub-second temporal precision for characterizing changes in patterns of activation over time. Increasingly fine anatomical resolution is available with functional magnetic resonance imaging (fMRI). However, fMRI is an indirect measure of neuronal electrical activity whose temporal resolution is too gross to resolve the rapidly shifting patterns of activity that are characteristic of actual neurophysiological processes. In contrast, electroencephalography (EEG) and event-related potential (ERP) methods have a temporal resolution typically in the one to five millisecond range, depending on the A/D rate. For simplicity the term EEG is used here in a general sense to refer both to recordings of brain electrical activity and, except where noted, to recordings of brain magnetic activity called magnetoencephalograms or MEGs. The nature of MEG recording technology and the relative strengths and weaknesses of EEG versus MEG approaches have been reviewed elsewhere (Cohen & Cuffin, 1991; Leahy et al., 1998; Williamson & Kaufman, 1987). From a broad perspective that considers all neuroimaging modalities, the differences between EEG and MEG are slight relative to their similarities. The sensitivity of the EEG to changes in mental activity has been recognized since Hans Berger reported a decrease in the amplitude of the dominant (alpha) rhythm of the EEG during mental arithmetic (Berger, 1929). In addition to the type of tonic alterations in brain electrical activity reported by Berger, EEG measurements of phasic, stimulus-related changes in brain activity (such as ERPs) are well-suited for measuring sub-second component processes of sensory, motor, and cognitive processes (Hillyard & Picton, 1987; Regan, 1989). Measurements of the coherence, correlation, or covariance of EEG time series from different electrode sites help generate hypotheses about the functional networks that form between different cortical regions during these processes. The temporal resolution and sensitivity of the EEG make it an ideal complement to fMRI. However, the spatial detail obtained in most EEG studies has been so coarse that it has only been possible to meaningfully interpret EEGs with respect to underlying functional neuroanatomy at the level of entire cortical lobes, if at all. The overriding limitations in this regard are that the EEG is most often recorded at a small number of scalp sites and spatial deblurring methods are not used to compensate for distortion due to conduction through the skull. Thus, even though the ability to infer the three-dimensional distribution of electrical sources in the brain from scalp EEG recordings has fundamental physical limits, the amount of spatial detail that can be gleaned from the scalp-recorded EEGs is often not appreciated.

2.   INCREASING SPATIAL DETAIL

Recording from more electrodes is the first requirement for extracting more spatial detailed from scalp-recorded EEGs. The nineteen-channel "10/20" montage of electrode placement commonly used in clinical EEG recordings has an inter-electrode distance of about 6 cm on a typical adult head (Jasper, 1958). This spacing may be sufficient for detecting signs of gross pathology or for differentiating the overall topography of ERP components, but it is insufficient for resolving finer grained topographical details that may be of importance in studying cognition. By increasing the number of electrodes to over 100, average inter-electrode distances of about 2.5 cm can be obtained on a typical adult head. This distance is within the the typical cortex-to-scalp point spread function—i.e., the size of the scalp representation of a small, discrete cortical source (Gevins et al., 1990).

For electrical (but not magnetic) recordings, the usefulness of such increased spatial sampling remains limited by the distortion of neuronal potentials as they are passively conducted through the highly resistive skull (Gevins et al., 1991; van den Broek et al., 1998). This distortion reflects a spatial low-pass filtering, which causes a blurring of the potential distribution at the scalp. In recent years a number of spatial enhancement methods have been developed for reducing this distortion.

The simplest and most widely used of these methods is the spatial Laplacian operator, usually referred to as the Laplacian Derivation (LD). It is computed as the second derivative in space of the potential field at each electrode. The LD is proportional to the current entering and exiting the scalp at each electrode site (Nunez, 1981; Nunez & Pilgreen, 1991), and is independent of the location of the reference electrode used for recording. The LD is relatively insensitive to signals that are common to the local group of electrodes used in its computation and is, therefore, relatively more sensitive to high spatial frequency local cortical potentials. A simple method of computing the LD assumes that electrodes are equidistant and at right angles to each other, an approximation that is only reasonable at a few scalp locations such as the vertex. A more accurate approach is based on measuring the actual three-dimensional position of the electrodes and using 3D spline functions to compute the LD over the actual shape of a subject's head (Le et al., 1994). The main shortcoming of the LD is that it unrealistically assumes that the skull has the same thickness and conductivity everywhere, which limits the improvement in spatial detail that the method can achieve.

This shortcoming of the LD can be ameliorated by using a realistic model of each subject's head to locally correct the EEG potential field for distortion resulting from conduction to the scalp. One such method is called "Finite Element Deblurring." It provides a computational estimate of the electrical potential field near the cortical surface by using a realistic mathematical model of volume conduction through the skull and scalp to downwardly project scalp-recorded signals (Gevins et al., 1991, 1994; Le & Gevins, 1993). Each subject's MRI is used to construct a realistic model of his or her head in the form of many small tetrahedral elements representing the tissues of scalp, skull, and brain. By assigning each tissue a conductivity value, the potential at all finite element vertices can be calculated using Poisson's equation. Given that the actual conductivity value of each of these finite elements is unknown, a constant value is used for the ratio of scalp to skull conductivity; the conductivity of each finite element is set by multiplying this constant by the local tissue thickness as determined from the MRI. Thus, even though true local conductivity is unknown, the procedure is well behaved with respect to this source of uncertainty, because it successfully accounts for relative conductivity variation due to regional differences in scalp and skull thickness.

The Deblurring method has been shown to be reliable and more accurate than the LD (Gevins et al., 1991, 1994), an improvement that occurs at the expense of obtaining and processing each subject's structural MRI. Although Deblurring can substantially improve the spatial detail provided by scalp recorded EEG, it does not usually provide additional information about the location of generating sources. Nevertheless, the improved spatial detail facilitates formation of more specific hypotheses about the distribution of active cortical areas during a cognitive task. Blurring of brain signals by the skull can largely be avoided by recording the magnetic rather than the electrical fields of the brain, because the skull has no effect on magnetic field topography. However, this transparency does not eliminate the need for utilizing a high density of sensors to accurately map the spatial topography of brain magnetic fields. Furthermore, the problems of localizing generator sources are equally severe for MEG as they are for EEG (see below). Further, the cost of MEG technology is more than an order of magnitude greater than that required for EEG studies, and the associated infrastructure required to perform MEG studies is more complex and inflexible. Thus, for most laboratories and for some applications, such as those in which a subject's head cannot be immobilized for long term monitoring or ambulatory recordings, MEG does not provide a viable alternative to EEG recordings.

3.   VERIFICATION AND APPLICATION OF HIGH RESOLUTION EEG IN EXPERIMENTAL STUDIES

Exploratory studies of Deblurring and other high resolution EEG techniques focused on the spatial enhancement of sensory ERPs, where a great deal of a priori knowledge exists concerning their underlying neural generators (Gevins et al., 1994). These studies demonstrated that Deblurred somatosensory responses isolated activity to the region of the central sulcus. Similar localization is obtained with movement-related potentials. Figure 1 illustrates the results of Deblurring potentials locked in time to a button press response made with the right hand. The major foci of activity occur in the contralateral (left hemisphere) in the somatomotor region of the pre- and post-central gyri. Demonstrations like these help verify the reasonableness of the approach. A better validation is obtained by comparison of the Deblurred potentials with subdural grid recordings in epileptic patients undergoing evaluation for ablative surgery. To date these validation studies have produced a reasonable degree of agreement between the Deblurred potentials and those measured directly at the cortical surface (Gevins et al., 1994).

Recent developments suggest that high-resolution EEG methods are useful tools in the experimental analysis of higher-order brain functions, and functional localization of cognitive processes inferred from spatially enhanced and anatomically registered neurophysiological measurements can be compared with the results of lesion studies and other neuroimaging techniques. As a complement to these approaches, the fine-grain temporal resolution of ERP measurements, in combination with improved topographic detail, adds valuable insights gained by characterizing both the regionalization of functions and the sub-second dynamics of their engagement. For example, spatial enhancement of EEGs related to component processes in reading has yielded results that are highly consistent with current knowledge of the functional neuroanatomy thought to be involved with visual pattern recognition and language functions (Gevins et al., 1995).

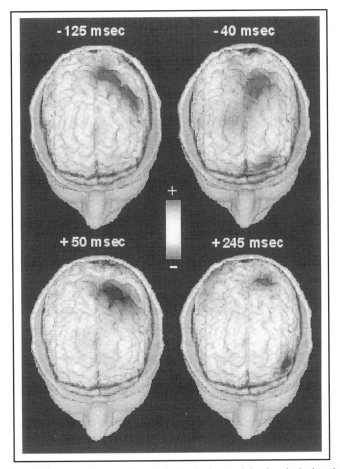

Figure 1. Deblurred Movement-Related Potentials. Evoked potentials, time-locked to the onset of a button press response made with the right middle finger, were recorded from a high density (128 channel) electrode montage attached to the scalp. The blurring effect of the scalp and skull were mathematically removed using the Deblurring method, and the resulting spatially sharpened data were projected onto the cortical surface, which was constructed from the subject's MRI. Activity at two instants prior to the button press (top) and two instants after the button press (bottom) are plotted. Both pre-and post-response activation are strongly lateralized to the left hemisphere–the hemisphere contralateral to the hand used to make the response. This figure shows lateralized activation of the precentral gyrus prior to and immediately after a button press response is made. Approximately 250 milliseconds after the response, the focus of activation moves to the postcentral gyrus (after Gevins, et al., 1999).

Modern EEG methods have also been used to study sub-second and multi-second distributed neural processes associated with working memory, the cognitive function of creating a temporary internal representation of information during focused thought (Gevins et al., 1996; McEvoy et al., 1998). In task conditions that placed a high load on working memory functions, subjects were asked to decide if the stimulus on each trial matched either the verbal identity or the spatial location of a stimulus occurring three trials (13.5 second) previously. This required subjects to concentrate on maintaining a sequence of three letter names or three spatial locations concurrently; they had to update that sequence on each trial by remembering the most recent stimulus and could drop the stimulus from four trials back. In two corresponding control conditions, only the verbal identity or spatial location of the first stimulus had to be remembered. Both spatial and verbal working memory tasks produced highly localized momentary modulation of ERPs over prefrontal cortical areas relative to control conditions, with Deblurred voltage maxima approximately over Brodmann's areas 9, 45, and 46 (Figure 2). These brief (~50 and to 200 milliseconds) events occurred in parallel with a sustained ERP wave, maximal over the superior parietal lobe and the supramarginal gyrus, with a slight right-hemisphere predominance. It began ~200 milliseconds after stimulus onset, returned to near baseline by ~600 milliseconds post-stimulus in control conditions and was sustained up to ~1 second or longer in the WM conditions. The sub-second ERP effects occurred in conjunction with multi-second changes in the ongoing EEG, of which the theta band power focused over midline frontal cortex is shown in Figure 2 (Gevins et al., 1997; Smith et al., 1999). These EEG findings may provide the first direct evidence in a single experiment supporting the idea that the various types of attention are associated with neural processes with distinct time courses in distinct neuronal populations. The increased theta band power may be a marker of the continuous focused attention required to perform the task and reflect engagement of the anterior cingulate gyrus, a conjecture supported by dipole modeling (Gevins et al., 1997). In contrast, the momentary attention required for scanning and updating the representations of working memory may be indexed by increased ERP amplitude peaks over lateralized regions of dorsolateral prefrontal cortex, while maintenance of a representation of the stimuli being remembered may be reflected in the parietally maximal ERP wave and other concomitant changes in the EEG (Gevins et al., 1996, 1997).

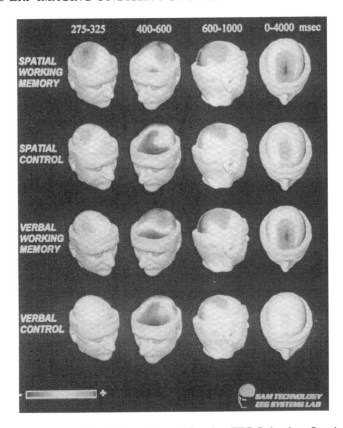

Figure 2. Deblurred Event-Related Potentials and Ongoing EEG Related to Sustained Focused Attention. High resolution EEG methods have made it possible to simultaneously measure both sub-second phasic and multi-second tonic regional brain activity during performance of cognitive tasks. In this experiment a sequence of increased sub-second ERP peaks and waves was observed over frontal (first and second columns) and parietal (third column) cortices during a difficult working memory task, in comparison to control conditions with lower working memory requirements. These sub-second changes in the working memory tasks were accompanied by longer lasting (4 second) increases in ongoing EEG theta band power (rightmost column). These EEG findings suggest that various types of attention are associated with neural processes that have distinct time courses in distinct neuronal populations. Amplitude scale is constant across experimental conditions within each column; ERP scale is voltage, EEG scale is z-scored spectral power (after Gevins et al., 1995).

These EEG and ERP measurements of sustained focused attention and working memory have high test-retest reliability (McEvoy et al., 2000), and vary in predictable ways across the lifespan (McEvoy et al., 2001). They also can be highly predictive of individual differences in cognitive ability and cognitive style as defined by traditional psychometric instruments (Gevins & Smith, 2000), and are also sensitive to transient cognitive impairments that can be produced by fatigue or some medications (Gevins & Smith, 1999; Gevins et al., 2001). Such measures might therefore serve an important role in clinical assessment techniques that incorporate both behavioural and neurophysiological indices.

4.   IDENTIFYING THE GENERATORS OF EEG

Neither the Laplacian Derivation, nor more advanced EEG spatial enhancement algorithms such as Deblurring, nor MEG recordings, provide any conclusive three-dimensional information about where the source of a scalp-recorded signal lies in the brain. In some cases, such as when healthy subjects perform difficult cognitive tasks and strong signals are recorded over areas of association cortex (i.e., dorsolateral prefrontal, superior and inferior parietal, inferotemporal and lateral temporal), the hypothesis that EEG potentials are generated in these areas is the most plausible. However, counter-examples can always be presented. In addition to visual examination of the potential field distribution, "dipole modeling" provides another method for generating hypotheses concerning the neuroanatomical loci responsible for generating neuroelectric events measured at the scalp (Fender, 1987; Scherg & Von Cramon, 1985). Dipole modeling uses iterative numerical methods to fit a mathematical representation of a focal, dipolar current source, or collection of such sources, to an observed scalp-recorded EEG or MEG field.

Source modeling does not, in general, provide a unique or necessarily physically correct answer about where in the brain activity recorded at the scalp is generated. This is so because solving for the source of an EEG or MEG distribution recorded at the scalp is a mathematically ill-conditioned "inverse problem" that has no unique solution; additional information and/or assumptions are required to choose among candidate source models. Although some of this a priori information is obvious (i.e., that the potentials must arise from the space occupied by the brain), other assumptions border on presupposing unknown information (i.e., that the potentials arise only from the cortex or that the number of active cortical areas is known).

One simple, convenient, and potentially clinically useful approach for potentials elicited by simple sensory stimulation is to assume that the scalp potential pattern arises from a single point dipole source (see Figure 3). Although not anatomically or physiologically realistic, such simple models can sometimes be useful for locating the center of mass of primary sensory cortex and hence major functional landmarks such as the central sulcus. When justified by simple voltage topography (e.g., Figure 4), models of this sort can also be useful for generating initial hypotheses about the possible sources underlying other phenomena.

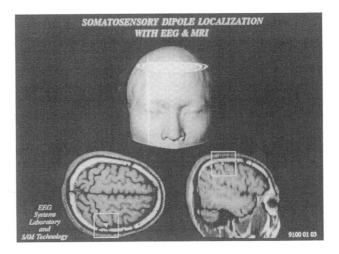

Figure 3. Localization of an EEG dipole model in the somatosensory cortex of the right hemisphere from scalp-recorded data evoked in response to transient electrical stimulation of the left index finger. This popular type of source generator localization modeling produces anatomically plausible results in the case of simple sensory stimulation (after Gevins, et al., 1993).

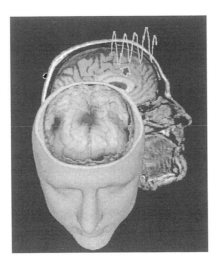

Figure 4. Deblurred frontal midline theta EEG activity and localization of corresponding source model in the region of the anterior cingulate cortex. Topographic data correspond to the difficult working memory task condition depicted in Figure 2 (Gevins et al., 1997). The data were processed with the Deblurring method, and the spatially sharpened results were projected onto the cortical surface, which was constructed from the subject's MRI. The upwards-oriented arrow superimposed on the midline sagittal image depicts the localization of a point dipole source model for these data (after Gevins, et al., 1999).

Most complex scalp-recorded neurophysiological phenomena are poorly approximated by a single dipole source model. Obtaining estimates of the strength and 3D locations of the underlying neuronal generators when there are multiple, time-overlapped active sources has widely recognized practical and theoretical difficulties (Miltner et al., 1994). An intensive effort is being allocated to the development of improved methods for source analysis for electrical phenomenon that are likely to arise from multiple and/or distributed sources (Gorodnitsky et al., 1995; Grave de Peralta-Menendez & Gonzalez-Andino, 1998; Koles, 1998; Tesche et al., 1995; Wang et al., 1993). Even so, regardless of which method is used to formulate them, such source generator hypotheses must ultimately be independently verified. In rare cases, this might be done in patient populations in the context of invasive recordings performed for clinical diagnostic purposes. More commonly, another type of imaging modality, such as fMRI, has to be employed. One promising approach to this issue is to use information about the cortical regions activated by a task as mapped by fMRI to constrain source models, and to derive information about the spatiotemporal dynamics of those sources from ERP measurements (George et al., 1995; Heinze et al., 1994; Mangun et al., 1998; Sereno, 1998; Simpson et al., 1995).

5.   DISTRIBUTED FUNCTIONAL NETWORKS OF SIMPLE COGNITIVE TASKS

Independently of whether definitive knowledge of source configurations exists, changes in the spatial distribution of EEG phenomena can be used to characterize the neural dynamics of thought processes. Even the simplest cognitive tasks require the functional coordination of a large number of widely distributed specialized brain systems. A simple response to a sensory stimulus involves the coordination of sensory, association and related areas that prepare for, register and analyze the stimulus, the motor systems that prepare for and execute the response, and other distributed neuronal networks. These distributed networks serve to allocate and direct attentional resources to the stimulus, to relate the stimulus to internal representations of the self and environment in order to decide what action to take, to initiate or inhibit the behavioral response, and to update internal representations after receiving feedback about the result of the action. In the ongoing EEG, hypotheses about functional interactions between cortical regions are sometimes drawn from measurements of statistical inter-relationships between time series recorded at different sites. These can be quantified by various measures of spectral, wave shape, or information-theoretic similarity, including: spectral coherence (Walter, 1963), correlation (Brazier & Casby, 1952; Gevins et al., 1981, 1983; Livanov, 1977), covariance (Gevins et al., 1987; Gevins et al., 1989a; Gevins et al., 1989b), information measures (Callaway & Harris, 1974; Mars & Lopes da Silva, 1987), nonlinear regression (Lopes da Silva et al., 1989) and multichannel time-varying autoregressive modeling (Gersch, 1987).

Some of the above methods can be used to characterize the spatiotemporal relationships of sub-second ERP components. Since the ERP waveform delineates the time course of event-related mass neural activity of a neuronal population, the coordination of two or more populations during task performance should be signaled by a consistent relationship between the morphology of the ERP waveforms emitted by these populations, with consistent time delay (Gevins & Bressler, 1988). If the relationships are linear, as they often appear to be, this coordinated activity might be measured by the lagged correlation or covariance between the ERPs, or segments of ERPs, from different regions (Gevins et al., 1987; Gevins et al., 1989a; Gevins et al., 1989b). One such measure of this type of process is referred to as an Event-Related Potential Covariance (ERPC). Of course, a significant covariance of this type is only a measure of statistical association, and does not map the actual neuronal pathways of interaction between functionally related populations. Studies of the neurogenesis of ERPCs are still in their infancy (Bressler et al., 1993; Gevins et al., 1994), and any interpretations of ERPCs in terms of the underlying neural processes that generate them must thus be made very cautiously. However, it is noteworthy that ERPC results to date have been highly consistent with the known large-scale functional neuroanatomy of frontal, parietal, and temporal association cortices. The ERPCs are beginning to provide fascinating glimpses of the complex, rapidly shifting distributed neuronal processes that underlie simple cognitive tasks.

The ERPC technique has yielded its most interesting results as a tool for studying preparatory attentional networks, the changes in brain activity associated with readiness for an impending event or action. For example, subjects in one experiment performed a task that required graded finger pressure responses with either the right or left hand proportional to visual numeric stimuli from 1 to 9 (Gevins et al., 1987; Gevins et al., 1989a; Gevins et al., 1989b). The hand to be used was cued one second before the stimulus. A 375-millisecond ERPC analysis window spanned the interval preceding the stimulus number in order to measure how ERP patterns differed according to the hand subjects expected to use. Figure 5 shows right-hand preparatory ERPCs for seven subjects for those trials for which the response (~0.5 to 1 second later) was subsequently either accurate or inaccurate. The set of subsequently accurate trials is characterized by covariances of the left prefrontal electrode with electrodes overlying the same motor, somatosensory and parietal areas that were involved in actual response executions (simultaneous measurement of flexor digitori muscle activity showed that the finger that would subsequently respond was not active during the preparatory interval). The preparatory patterns preceding inaccurate responses differed markedly from those preceding accurate responses, with fewer ERPCs between the left frontal site and other electrodes. Such results suggest that one important role of frontal lobe integrative mechanisms is the anticipatory scheduling and coordination of the activation of those specialized brain regions that will participate in an upcoming cognitive event.

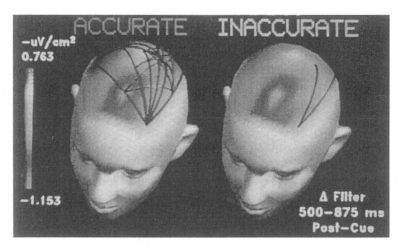

Figure 5. Preparatory Event-Related Potential Covariance (ERPC) patterns preceding accurate and inaccurate responses. ERPCs involving left frontal, midline precentral and left central and parietal electrode sites are prominent in patterns preceding (by 0.5 to 1 second) accurate responses (left). The number and magnitude of ERPCs are smaller preceding inaccurate responses (right) (after Gevins, et al., 1995).

Finally, an obvious but often unappreciated feature of EEG technology is worth mentioning, namely its extreme compactness and simplicity. This fact has important practical considerations, which frequently fail to be considered in scientific discussions of brain mapping methods. For example, the EEG has the potential to serve as a sensitive, low-cost, and portable monitor of cognition for clinical assessment and other applications (Gevins, 1998; Gevins et al., 1998). The compactness of EEG technology also means that, unlike all other functional neuroimaging modalities (which require massive machinery, large teams of technicians, and complete immobilization of the subject) EEGs can be collected from an ambulatory subject who is literally wearing the entire recording apparatus. This feature of EEGs will facilitate research into the as yet uncharted territory of how brains think when performing everyday activities in the real world (Gevins et al., 1995; Smith et al., 2001).

6.   CONCLUSIONS

The neurophysiology of mentation involves rapid coordination of processes in widely distributed cortical and subcortical areas. The electrical signals that accompany higher cognitive functions are subtle, spatially complex, and change both in a tonic multi-second fashion and phasically in sub-second intervals in response to environmental demands and internal representations of environment and self. No single brain imaging technology is currently capable of providing both near millimeter precision in localizing regions of activated tissue and sub-second temporal precision for characterizing changes in patterns of activation over time. However, by combining several technologies, it seems possible to achieve this fine degree of spatiotemporal resolution. Modern high-resolution EEG is especially well suited for monitoring rapidly changing regional patterns of neuronal activation accompanying purposive behaviors, while fMRI seems ideal for precisely determining their three-dimensional localization and distribution. Current research is seeking to determine how to combine EEG and fMRI data from the same subjects doing the same tasks.

ACKNOWLEDGMENTS

We thank our many colleagues at the San Francisco Brain Research Institute (formerly EEG Systems Laboratory) and SAM Technology, past and present, for their contributions to the work described here. This research was supported by grants from The Air Force Office of Scientific Research, The National Institute of Mental Health, The National Institute of Neurological Disorders and Stroke, The National Science Foundation, The National Aeronautics and Space Administration, The Air Force Research Laboratory, The Office of Naval Research, The National Institute of Alcoholism and Alcohol Abuse, The National Institute of Drug Abuse, The National Institute of Child Health and Human Development and the National Institute of Aging of the United States Federal Government.

Figures 1 and 4 reprinted from Gevins, A., Smith, M.E., McEvoy, L.K., Leong, H. & Le, J. (1999). Electroencephalographic imaging of higher brain function. *Philosophical Transactions of the Royal Society of London, 354,* 1125-1133. Copyright (2002), with permission from The Royal Society.

Figures 2 and 5 reprinted from Gevins, A., Leong, H., & Smith, M.E., Le, J. & Du, R. (1995). Mapping cognitive brain function with modern high resolution electroencephalography. *Trends in Neurosciences, 18,* 427-461. Copyright (2002), with permission from Elsevier Science.

Figure 3 reprinted from Gevins, A., Le, J., Brickett, P., Reutter, B., & Desmond, J. (1993) Seeing through the skull: Advanced EEGs use MRIs to accurately measure cortical activity from the scalp. *Brain Topography, 4,* 125-131. Copyright (2002), with permission from Kluwer Academic Publishers.

REFERENCES

Berger, H. (1929). Uber das Elektroenzephalogramm des Menschen. *Archives of Psychiatry, 87* (Nervenk), 527-570.

Brazier, M.A.B., & Casby, J.U. (1952). Crosscorrelation and autocorrelation studies of electroencephalographic potentials. *Electroencephalography and Clinical Neurophysiology, 4,* 201-211.

Bressler, S.L., Coppola, R., & Nakamura, R. (1993). Episodic multiregional cortical coherence at multiple frequencies during visual task performance. *Nature, 366,* 153-155.

Callaway, E., & Harris, P. (1974). Coupling between cortical potentials from different areas. *Science, 183,* 873-875.

Cohen, D., & Cuffin, B.N. (1991). EEG versus MEG localization accuracy: Theory and experiment. *Brain Topography, 4,* 95-103.

Fender, D.H. (1987). Source localization of brain electrical activity. In A.S. Gevins & A. Rémond (Eds.), *Methods of analysis of brain electrical and magnetic signals. Vol. 1* (pp. 355-403). Amsterdam: Elsevier.

George, J.S., Aine, C.J., Mosher, J.C., Schmidt, D.M., Ranken, D.M., Schlitt, H.A., Wood, C.C., Lewine, J.D., Sanders, J.A., & Belliveau, J.W. (1995). Mapping function in the human brain with magnetoencephalography, anatomical magnetic resonance imaging, and functional magnetic resonance imaging. *Journal of Clinical Neurophysiology, 12,* 406-431.

Gersch, W. (1987). Non-stationary multichannel time series analysis. In A.S. Gevins & A. Rémond (Eds.), *Methods of analysis of brain electrical and magnetic signals. Vol. 1* (pp. 261-296). Amsterdam: Elsevier.

Gevins, A. (1998). The future of electroencephalography in assessing neurocognitive functioning. *Electroencephalograph and Clinical Neurophysiology, 106*, 165-172.

Gevins, A., Le, J., Brickett, P., Reutter, B., & Desmond, J. (1993) Seeing through the skull: Advanced EEGs use MRIs to accurately measure cortical activity from the scalp. *Brain Topography, 4,* 125-131.

Gevins, A., Leong, H., & Smith, M.E., Le, J. & Du, R. (1995). Mapping cognitive brain function with modern high resolution electroencephalography. *Trends in Neurosciences, 18,* 427-461.

Gevins, A., & Smith, M.E. (1999). Detecting transient cognitive impairment with EEG pattern recognition methods. *Aviation, Space, and Environmental Medicine, 70,* 1018-1024.

Gevins, A., & Smith, M.E. (2000). Neurophysiological measures of working memory and individual differences in cognitive ability and cognitive style. *Cerebral Cortex, 10,* 829-839.

Gevins, A., Smith, M.E., Leong, H., McEvoy, L., Whitfield, S., Du, R., & Rush, G. (1998). Monitoring working memory load during computer-based tasks with EEG pattern recognition methods. *Human Factors, 40,* 79-91.

Gevins, A., Smith, M.E., McEvoy, L., & Yu, D. (1997). High resolution EEG mapping of cortical activation related to working memory: effects of task difficulty, type of processing, and practice. *Cerebral Cortex, 7,* 374-385.

Gevins, A., Smith, M.E., & McEvoy, L.K. (2002). Tracking the cognitive pharmacodynamics of psychoactive substances with combinations of behavioural and neurophysiological measures. *Neuropsychopharmacology, 26,* 27-39.

Gevins, A., Smith, M.E., McEvoy, L.K., Leong, H. & Le, J. (1999). Electroencephalographic imaging of higher brain function. *Philosophical Transactions of the Royal Society of London, 354,* 1125-1133.

Gevins, A.S., & Bressler, S.L. (1988). Functional topography of the human brain. In G. Pfurtscheller (Ed.), *Functional brain imaging* (pp. 99-116). Toronto: Hans Huber Publishers.

Gevins, A.S., Bressler, S.L., Morgan, N.H., Cutillo, B.A., White, R.M., Greer, D., & Illes, J. (1989a). Event-related covariances during a bimanual visuomotor task. Part I. Methods and analysis of stimulus and response-locked data. *Electroencephalography and Clinical Neurophysiology, 74,* 58-75.

Gevins, A.S., Bressler, S.L., Morgan, N.H., Cutillo, B.A., White, R.M., Greer, D., & Illes, J. (1989b). Event-related covariances during a bimanual visuomotor task. Part II. Preparation and feedback. *Electroencephalography and Clinical Neurophysiology, 74,* 147-160.

Gevins, A.S., Brickett, P., Costales, B., Le, J., & Reutter, B. (1990). Beyond topographic mapping: Towards functional-anatomical imaging with 124-channel EEGs and 3-D MRIs. *Brain Topography, 3,* 53-64.

Gevins, A.S., Cutillo, B.A., Desmond, J.E., Ward, M., & Bressler, S.L. (1994). Subdural grid recordings of distributed neocortical networks involved with somatosensory discrimination. *Electroencephalography and Clinical Neurophysiology, 92,* 282-290.

Gevins, A.S., Cutillo, B.A., & Smith, M.E. (1995). Regional modulation of high resolution evoked potentials during verbal and nonverbal matching tasks. *Electroencephalography and Clinical Neurophysiology, 94,* 129-147.

Gevins, A.S., Doyle, J.C., Cutillo, B., Schaffer, R.E., Tannehill, R.S., Ghannam, J.H., Gilcrease, V.A., & Yeager, C.L. (1981). Electrical potentials in human brain during cognition: New method reveals dynamic patterns of correlation. *Science, 213,* 918-922.

Gevins, A.S., Le, J., Brickett, P., Reutter, B., & Desmond, J. (1991). Seeing through the skull: Advanced EEGs use MRIs to accurately measure cortical activity from the scalp. *Brain Topography, 4,* 125-131.

Gevins, A.S., Le, J., Martin, N.K., Reutter, B., Desmond, J., & Brickett, P. (1994). High resolution EEG: 124-channel recording, spatial deblurring and MRI integration methods. *Electroencephalograph and Clinical Neurophysiology, 90,* 337-358.

Gevins, A.S., Leong, H., Du, R., Smith, M.E., Le, J., DuRousseau, D., Zhang, J., & Libove, J. (1995). Towards measurement of brain function in operational environments. *Biological Psychology, 40,* 169-186.

Gevins, A.S., Morgan, N.H., Bressler, S.L., Cutillo, B.A., White, R.M., Illes, J., Greer, D.S., Doyle, J.C., & Zeitlin, G.M. (1987). Human neuroelectric patterns predict performance accuracy. *Science, 235,* 580-585.

Gevins, A.S., Schaffer, R.E., Doyle, J.C., Cutillo, B., Tannehill, R.S., & Bressler, S.L. (1983). Shadows of thought: Shifting lateralization of human brain electrical patterns during brief visuomotor task. *Science, 220,* 97-99.

Gevins, A.S., Smith, M.E., Le, J., Leong, H., Bennett, J., Martin, N., McEvoy, L., Du, R., & Whitfield, S. (1996). High resolution evoked potential imaging of the cortical dynamics of human working memory. *Electroencephalography and Clinical Neurophysiology, 98,* 327-348.

Gorodnitsky, I.F., George, J.S., & Rao, B.D. (1995). Neuromagnetic source imaging with FOCUSS: a recursive weighted minimum norm algorithm. *Electroencephalography and Clinical Neurophysiology, 95,* 231-251.

Grave de Peralta-Menendez, R., & Gonzalez-Andino, S.L. (1998). A critical analysis of linear inverse solutions to the neuroelectric inverse problem. *IEEE Transactions in Biomedical Engineering, 45,* 440-448.

Heinze, H.J., Mangun, G.R., Burchert, W., Hinrichs, H., Scholz, M., Munte, T.F., Gos, A., Scherg, M., Johannes, S., & Hundeshagen, H. (1994). Combined spatial and temporal imaging of brain activity during visual selective attention in humans. *Nature, 372,* 543-546.

Hillyard, S.A., & Picton, T.W. (1987). Electrophysiology of cognition. In V.B. Mountcastle & F. Plum (Eds.), *Handbook of Physiology: Vol. 5. Higher Functions of the Brain* (2nd ed., pp. 519-584). Bethesda, MD: American Physiological Society.

Jasper, H.H. (1958). The ten-twenty electrode system of the International Federation. *Electroencephalograph and Clinical Neurophysiology, 10,* 371-375.

Koles, Z.J. (1998). Trends in EEG source localization. *Electroencephalograph and Clinical Neurophysiology, 106,* 127-137.

Le, J., & Gevins, A.S. (1993). Method to reduce blur distortion from EEGs using a realistic head model. *IEEE Transactions on Biomedical Engineering, 40,* 517-528.

Le, J., Menon, V., & Gevins, A. (1994). Local estimate of the surface Laplacian derivation on a realistically shaped scalp surface and its performance on noisy data. *Electroencephalography and Clinical Neurophysiology, 92,* 433-441.

Leahy, R.M., Mosher, J.C., Spencer, M.E., Huang, M.X., & Lewine, J.D. (1998). A study of dipole localization accuracy for MEG and EEG using a human skull phantom. *Electroencephalography and Clinical Neurophysiology, 107,* 159-173.

Livanov, M.N. (1977). *Spatial organization of cerebral processes.* New York: Wiley.

Lopes da Silva, F., Pijn, J.P., & Boeijinga, P. (1989). Interdependence of EEG signals: Linear vs. nonlinear associations and the significance of time delays and phase shifts. *Brain Topography, 2,* 9-18.

Mangun, G.R., Buonocore, M.H., Girelli, M., & Jha, A.P. (1998). ERP and fMRI measures of visual spatial selective attention. *Human Brain Mapping, 6,* 383-389.

Mars, N.J., & Lopes da Silva, F.H. (1987). EEG analysis methods based on information theory. In A.S. Gevins & A. Rémond (Eds.), *Methods of Analysis of Brain Electrical and Magnetic Signals: Vol. 1* (pp. 297-307). Amsterdam: Elsevier.

McEvoy, L.K., Pellouchoud, E., Smith, M.E.S., & Gevins, A. (2001). Neurophysiological signals of working memory in normal aging. *Cognitive Brain Research, 11, 363-376.*

McEvoy, L.K., Smith, M.E., & Gevins, A. (1998). Dynamic cortical networks of verbal and spatial working memory: Effects of memory load and task practice. *Cerebral Cortex, 8,* 563-574.

McEvoy, L.K., Smith, M.E., & Gevins, A. (2000). Test-retest reliability of cognitive EEG. *Clinical Neurophysiology, 111,* 457-463.

Miltner, W., Braun, C., Johnson Jr., R., Simpson, G.V., & Ruchkin, D.S. (1994). A test of brain electrical source analysis (BESA): A simulation study. *Electroencephalography and Clinical Neurophysiology, 91,* 295-310.

Nunez, P.L. (1981). *Electric fields in the brain: The neurophysics of EEG.* New York: Oxford University Press.

Nunez, P.L., & Pilgreen, K.L. (1991). The spline-Laplacian in clinical neurophysiology: A method to improve EEG spatial resolution. *Journal of Clinical Neurophysiology, 8,* 397-413.

Regan, D. (1989). *Human brain electrophysiology.* New York: Elsevier.

Scherg, M., & Von Cramon, D. (1985). Two bilateral sources of the late AEP as identified by a spatio-temporal dipole model. *Electroencephalography and Clinical Neurophysiology, 62,* 32-44.

Sereno, M.I. (1998). Brain mapping in animals and humans. *Current Opinion in Neurobiology, 8*, 188-194.

Simpson, G.V., Pflieger, M.E., Foxe, J.J., Ahlfors, S.P., Vaughan, J.H.G., Hrabe, J., Ilmoniemi, R.J., & Lantos, G. (1995). Dynamic neuroimaging of brain function. *Journal of Clinical Neurophysiology, 12*, 432-449.

Smith, M.E., Gevins, A., Brown, H., Karnik, A., & Du, R. (2001). Monitoring task load with multivariate EEG measures during complex forms of human computer interaction. *Human Factors, 43, 366-380.*

Smith, M.E., McEvoy, L.K., & Gevins, A. (1999). Neurophysiological indices of strategy development and skill acquisition. *Cognitive Brain Research, 7*, 389-404.

Tesche, C.D., Uusitalo, M.A., Ilmoniemi, R.J., Huotilainen, M., Kajola, M., & Salonen, O. (1995). Signal-space projections of MEG data characterize both distributed and well-localized neuronal sources. *Electroencephalograph and Clinical Neurophysiology, 95*, 189-200.

van den Broek, S.P., Reindeers, F., Donderwinkel, M., & Peters, M.J. (1998). Volume conductions effects in EEG and MEG. *Electroencephalography and Clinical Neurophysiology, 106*, 522-534.

Walter, D.O. (1963). Spectral analysis for electroencephalograms: Mathematical determination of neurophysiological relationships from records of limited duration. *Experimental Neurology, 8*, 155-181.

Wang, J.Z., Williamson, S.J., & Kaufman, L. (1993). Magnetic source imaging based on the minimum-norm least-squares inverse. *Brain Topography, 5*, 365-371.

Williamson, S.J., & Kaufman, L. (1987). In A.S. Gevins & A. Remond (Eds.), *Handbook of electroencephalography and clinical neurophysiology, Vol. 1: Methods of analysis of brain electrical and magnetic signals* (pp. 405-448). Amsterdam: Elsevier.

Chapter 9

EEG THETA, MEMORY, AND SLEEP

WOLFGANG KLIMESCH
Department of Physiological Psychology, University of Salzburg, Austria

1.   INTRODUCTION

In contrast to the alpha rhythm, which is the dominant large scale oscillation in the human EEG, most of what is known about the theta rhythm stems from animal research. Therefore, the most important properties and the functional meaning of the hippocampal EEG will be reviewed first. Research with human subjects then will be reported, which indicates that an event-related increase in theta power is associated with increasing memory demands in a similar way as was found in animal research for the hippocampal EEG. Finally, it will be shown that sleep is important for memory consolidation. The involvement of the hippocampal formation will be discussed.

2.   HIPPOCAMPAL THETA: SOME BASIC FACTS

The frequency of theta recorded from the hippocampus of lower mammals can vary between about 3.5 to 12 Hz (Lopes da Silva, 1992) and, therefore, shows a much wider frequency range than in humans where theta lies within a range of about 4 to 7.5 Hz. A regular oscillatory pattern can be observed in the hippocampal EEG (which also is termed Rhythmic Slow Activity or RSA) if animals make voluntary movements (Vanderwolf, 1992; Vanderwolf & Robinson, 1981), during exploratory behavior (Buzsaki et al., 1994) and also in REM sleep (Winson, 1990). In the absence of exploratory behavior (e.g., in slow wave sleep, SWS, or during alert immobility) the hippocampal EEG shows slow irregular activity in the delta frequency range (of about 1.5 to 4 Hz), which has received different names (Irregular Slow Activity, ISA; Large Irregular Activity, LIA; Sharp Waves, SPW) but will be called here SPW (Buzsaki et al., 1994).

Theta is usually recorded from microelectrodes implanted in the CA1 or the dentate layers of the hippocampus. It is induced from the septum, which serves as a primary pacemaker (Petsche et al., 1962; for reviews see Buzsaki et al., 1983 and Miller, 1991). The extracellular currents that underlie theta activity reflect synchronous fluctuations in the membrane potentials of pyramidal and granule cells (Bland, 1986, Buzsaki et al., 1983). Two facts are of crucial importance: (1) Most excitatory (pyramidal, granula) cells are silent during theta activity, only interneurons are bursting in theta (and gamma) frequency, paced by septal cells. (2) Theta operates to silence most principal cells (pyramidal, granule) by keeping their membrane voltage below but at the same time close to the firing threshold.

During theta activity—induced by septal neurons—a few entorhinal afferents and/or granula cells are sufficient to selectively activate principal cells that start bursting in the theta frequency range. When theta activity ceases, SPWs appear. Although there is high overlap, some brain regions are differentially involved in theta and SPW activity as shown schematically in Figure 1. Note that superficial layers of the entorhinal cortex that are an important input structure for the hippocampus are already paced within theta frequency. Deep layers that send fibers to other parts of the neocortex display SPWs after theta ceases. It has been emphasized that SPWs develop from the cells that fire in synchrony with the theta rhythm (Buzaki, 1996).

2.1   HIPPOCAMPAL ACTIVITY AND BEHAVIOR

It is well known that the frequency of theta is related to different types of behavior. Voluntary movements that are observed during exploratory behavior (e.g., walking, jumping, etc.) can be characterized by a highly regular theta oscillation, which is termed type 1 theta. In rodents, type 1 theta ranges from about 6.5 to 12 Hz. A somewhat slower and more irregular theta frequency can be observed during a state of immobile alertness, if sensory stimuli are presented while animals are immobile but alert and, thus, in a state of arousal (Montoya et al., 1989). This type 2 theta frequency varies within a range of about 4-9 Hz. A third type of behavior characterized by intermittent SPWs (with a duration of about 40 to 120 milliseconds and a frequency of about 0.02 to 3.5 Hz) can be observed during sleep and "automatic" motor patterns. Thus, frequency and regularity of hippocampal activity distinguish among different types of behavioral states such as slow wave sleep (SWS), exploratory behavior, and awake immobility (Buzsaki et al., 1983).

Shaded regions typically display theta activity (during exploratory behavior and REM sleep

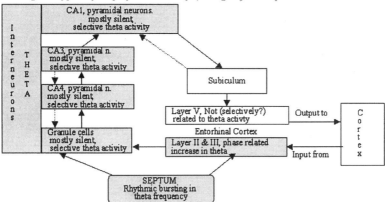

When theta oscillations disappear, SPW's (sharp waves) appear (shaded areas), during automatic movements and SWS sleep

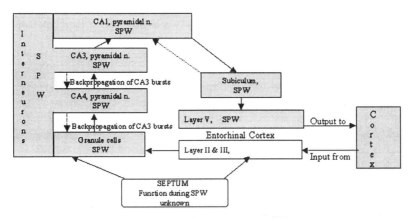

Figure 1. Interplay between theta and sharp waves (SPWs, in the delta frequency range). Theta is an inhibitory rhythm that operates to filter out highly selective excitatory network connections (Figure 1a). These connections are assumed to form the neuronal basis for the encoding of new (explicit) information. When theta ceases, they form the basis for the appearance of SPWs, which may be the electrophysiological correlate of memory consolidation (after Buzsaki, 1996).

Attentional demands are absent during SWS (when hippocampal activity is slowest and completely irregular) but highest during exploratory behavior (when theta frequency is highest and most regular). It is reasonable to assume that attention increases from SWS, awake immobility, automatic motor patterns (such as drinking, eating, face washing, and grooming), high alertness during immobility, and finally to voluntary movements during exploratory behavior. Hence, a perfect association between the frequency of the hippocampal EEG (within a range of 0.02-12 Hz) and attentional demands can be detected. In addition, with increasing attention, the hippocampal EEG becomes more regular until EEG-theta becomes synchronized within the small range of peak theta frequency (cf. Leung et al., 1982; Buzsaki et al., 1983).

Theta phase as well as theta power and frequency are related to behavior. Within the pyramidal cell layer of the rat hippocampus, cells have been found that unlike theta cells respond with a more complex pattern of spikes. O'Keefe and Dostrovsky (1971) observed that this cell type responds to the animal's location in the environment. Phase locked to the concurrent EEG-theta, these 'place cells' fire several bursts of spikes as the rat runs through a particular location in its environment, which is called the place field of that cell. The most important result is that the firing of place cells began at a particular phase of theta frequency as the rat entered the place field. Within a particular place field the angle of theta phase was related to a particular spatial location (O'Keefe & Recce, 1993; O'Keefe, 1993). Thus, when considering the place field a specific stimulus, theta frequency is phase locked to the appearance of this stimulus (cf. Givens, 1996).

2.2   HIPPOCAMPAL THETA REFLECTS MEMORY PROCESSES

The hippocampal EEG distinguishes between different behavioral states, but what exactly is the functional meaning of the different "types" of theta? A first hint comes from the fact that type 1 theta is associated with a behavioral state (i.e., exploratory behavior) in which not only attentional but also memory demands are highest. It is a well established finding that the hippocampal formation together with a complex structure of other brain regions (Squire, 1992) is important for the encoding and possibly also retrieval of new information (Scoville & Milner, 1957). Hence, hippocampal theta rhythm might be related to these types of memory processes. Strong evidence for this hypothesis has come from studies that have documented a preference for long-term potentiation (LTP) to occur in the hippocampal formation, and that theta activity induces or at least enhances LTP (Larson et al., 1986, Greenstein et al., 1988; Maren et al., 1994). Because LTP is considered the most important electrophysiological memory mechanism for the encoding of new information, experimental evidence for a functional relation between LTP and theta provides an important argument supporting theta as an electrophysiological correlate of working memory. Indeed, the induction of LTP is optimal with stimulation patterns that mimic the theta rhythm (Larson et al., 1986), whereas not repeated and short stimulations are usually ineffective. In addition, the induction of LTP depends on the phase of theta rhythm, and the strength of the induced LTP was found to increase linearly with increasing theta power (Maren et al., 1994, Figure 6). These findings provide consistent evidence that hippocampal theta is related to the encoding of new information, just as LTP is.

Theories and reviews about hippocampal theta and memory are available with respect to the question whether theta activity can be analyzed in the human scalp EEG (Buzsaki et al., 1994; Lisman & Idiart, 1995; O'Keefe & Burgess, 1999). Miller's (1991) theory of resonant loops is of particular importance. He assumes a rhythmic interplay between the hippocampus and cortex. The human scalp EEG primarily reflects cortical activity. However, theta frequency in the hippocampus, deep inside the brain, would be biophysically difficult to detect from scalp electrodes. If hippocampal theta is induced into the cortex, given the above, it should be possible to detect memory related changes in theta activity even in the human scalp EEG.

3.   THETA ACTIVITY IN THE HUMAN SCALP EEG

The traditional view holds that theta frequency lies within a fixed range of about 4 to 8 Hz just below alpha frequency (about 8-13 Hz). A fixed frequency band would be justified only if it can be demonstrated that theta does not vary among subjects. Thus, before considering findings from the human scalp EEG it is important to address the following two questions: (1) Is there a physiological criterion that specifies which frequency marks the transition between alpha and theta? (2) Does theta frequency vary in a similar way as has been demonstrated for alpha frequency (Klimesch, 1999)?

The answer to the first question is surprisingly easy. It is a well established fact that with increasing task demands theta synchronizes, whereas alpha desynchronizes. If EEG power in a resting condition is compared with a test condition, alpha power becomes suppressed (desynchronizes), whereas theta power increases (synchronizes). The specific frequency in the power spectrum that marks the transition between an event-related increase in theta and a decrease in alpha power can be considered the individual transition frequency (TF) between the alpha and theta band for individual subjects.

When using this method to estimate TF, the second question can also be answered: TF shows a large interindividual variability (ranging from about 4 to 7 Hz), which is linked to the interindividual variability of alpha peak frequency. Preliminary evidence for a covariation between theta and alpha frequency was already found by Klimesch et al. (1994). Additional evidence for such a covariation has also been obtained using the following method (Klimesch et al., 1996). First, power spectra for the reference and test intervals were calculated for each subject and averaged over all trials and all leads. Then, the frequency of the transition region between the theta and alpha band was determined within a frequency window of 3.5-7.5 Hz. For those few subjects, who showed an asymmetric alpha desynchronization (with no desynchronization in the lower alpha) and therefore failed to show an intersection (in the range of 3.5-7.5 Hz), the transition between the theta and alpha band was considered that frequency where the difference between the test and the reference interval reached a minimum. Klimesch et al. (1996) found that Spearman's rank correlation between alpha peak frequency and TF yielded a significant value of rho=0.64 (p<.02), with similar results reported by Doppelmayr et al. (1998a). These findings document that theta varies as a function of alpha frequency and suggest to use individual alpha frequency as a common reference point for defining different frequency bands including theta. For theta, the individual determination of frequency bands may even be more important, because the effects of theta synchronization are otherwise masked by alpha desynchronization particularly in the range of the TF.

In general, TF lies about 4 Hz below the individually determined alpha frequency. As an example, in a sample of 10 subjects with a mean alpha frequency of 10.7 Hz, TF lies at 6.7 Hz (Klimesch et al., 1996). If a bandwidth of 4 Hz would be assumed, theta frequency would cover a range of 2.7 to 6.7 Hz. This produces an estimate that appears too low for the lower frequency boundary of 2.7 Hz since this frequency is considered to belong to the delta frequency range. To avoid overlap with delta, theta frequency can be defined as that band with a width of just 2 Hz that falls below TF. Thus, in our example theta is a band of 4.7 to 6.7 Hz. All of the findings reported below are based on individually adjusted theta bands of 2 Hz width—results that suggest the theta frequency range in humans may be much smaller than originally assumed.

3.1   THETA ACTIVITY IN THE HUMAN SCALP EEG AND MEMORY PROCESSES

A well known procedure to measure event-related changes in band power is based on a method originally proposed by Pfurtscheller and Aranibar (1977). Event-related band power changes are defined as the percentage of a decrease termed event-related desynchronization (ERD) or increase termed event-related synchronization (ERS) in band power during a test with respect to a reference interval. The measurement of ERD/ERS is done in several steps. First, the EEG is band pass filtered within defined frequency bands, with the filtered data squared and then averaged within consecutive time intervals (e.g., 125 milliseconds). Second, the obtained data are averaged over the number of epochs. Third, band power changes are expressed as the percentage of a decrease or increase in band power during a test as compared to a reference interval by using the following formula: ERD = ((band power reference - band power test)/(band power reference))*100. Note that desynchronization is reflected by positive ERD values, whereas synchronization is reflected by positive ERS (or negative ERD) values.

The studies summarized below were designed to investigate the question of whether an event-related increase in theta power selectively reflects the successful encoding and/or retrieval of new episodic information. This hypothesis was derived from reviews focusing on findings in functional neuroanatomy, electrophysiology, amnesia, and memory research; it was proposed that the hippocampal theta rhythm primarily reflects the encoding and retrieval of episodic memory (Klimesch, 1995, 1996, 1999).

The fact that theta power increases in a large variety of different tasks seems to contradict the suggested hypothesis of a specific relationship between theta and the processing of new information (Arnolds et al., 1980; Schacter, 1977). However, the processing of new information is in some way and extent necessary for the performance of almost any type of task. Of particular importance is the experimental control of unspecific factors (such as attentional demands, task difficulty, and cognitive load) that usually accompany the processing of new information. The following studies compared different types of memory demands (such as episodic and semantic memory) and processes (such as incidental and episodic encoding, successful and unsuccessful encoding and/or retrieval).

In a first study, the relationship between theta synchronization, alpha desynchronization, episodic and semantic memory demands was investigated (Klimesch et al., 1994). The experimental design consisted of two parts. Subjects first performed a semantic congruency task in which they judged whether the sequentially presented words of concept-feature pairs (such as "eagle-claws" or "pea-huge") were congruent. They were then asked to perform an unexpected episodic recognition task. This approach helped to prevent subjects from using semantic encoding strategies and to increase episodic memory demands. In the episodic task, the same word pairs were presented together with new distractors (generated by re-pairing the already known concept-feature words). In order to perform the task correctly, subjects had to know whether or not a particular concept-feature pair was already presented during the semantic task. Because distractors were semantically similar and generated by re-pairing previously presented words, subjects were able to give a correct response only if new episodic information (represented by a specific combination of a concept and feature word) was actually stored in memory. The results of a reaction time study with the same experimental paradigm (Kroll & Klimesch, 1992, Experiment 4) indicated that semantic features speeded up semantic but slowed down episodic decision times. Thus, semantic and episodic memory processes can well be differentiated behaviorally (for a review, see Klimesch, 1994).

Furthermore, as pairs of items are presented, the episodic and semantic task could be performed only after the second item of a pair (i.e., the feature) was presented. The critical electrophysiological issue was to compare the amount of theta synchronization during the presentation of the concept and feature word in the episodic and semantic task. Only correctly identified concept feature pairs were analyzed. The results demonstrated that only in the episodic task and only when the feature word was processed was the expected increase in theta power observed. During the processing of the concept words, theta power evinced almost identical values in the semantic as well as the episodic task. Because the same words were presented in both tasks and all other variables were kept constant (exposure time, length of interstimulus interval, etc.), the findings support the hypothesis of a specific relationship between memory (for new episodic information) and theta synchronization. The fact that the theta band responds selectively to episodic task demands is also demonstrated by the finding that the lower and the upper alpha band show quite different results. Whereas event-related changes in the lower alpha band were comparatively small, the upper alpha band shows a much larger degree of desynchronization during the semantic as compared the episodic task. Thus, there is dissociation between theta synchronization, which is maximal during the processing of new information, and upper alpha desynchronization, which is maximal during the processing of semantic information.

Similar results were obtained in a study with three experiments in which words were presented and had to be encoded into working memory (Klimesch et al., 1997b). In all experiments only at occipital sites and only during the first 500 milliseconds after a word was presented visually, a short-lasting theta synchronization was observed. This increase in theta power was particularly strong at occipital areas and most likely reflects the encoding of new information at occipital sites. In a replication and extension of these findings, it was also demonstrated that the extent of upper alpha desynchronization is significantly correlated with semantic memory performance, whereas the extent of theta synchronization is significantly correlated with episodic memory performance (Klimesch et al., 1997a). In summary, in the theta band, episodic memory performance is associated with an event-related increase in power, whereas the opposite holds true for semantic memory and the upper alpha band (see also Klimesch et al., 2000).

An additional prediction of the theta hypothesis is that during the encoding of those words that will later be remembered (e.g., in a recognition task), a significantly stronger theta synchronization is expected compared to words that cannot be remembered later. Another prediction is that during the actual recognition process, correctly recognized targets will show a significantly stronger increase in theta power compared to not recognized targets or distractors. These predictions were tested in another study (Klimesch et al., 1997c) in which a set of 96 words was used as targets, with sub-samples of 16 words each comprised of one of 6 categories (birds, fruits, vegetables, vehicles, clothes, and weapons). The 96 distractors were selected such that for each target (e.g., robin) a semantically similar distractor (e.g., sparrow) was presented, so that just as for the targets, the 96 distractors are subdivided into the same 6 semantic categories, each comprising 16 words. This similarity between distractors and targets guarantees that subjects must rely on episodic information rather than, for example, semantic familiarity to make a correct decision in the recognition task (e.g. if a subject has to distinguish the target word "robin" from the distractor "sparrow", semantic information representing the meaning of the words will not be helpful). For a correct response, the subject has to remember which of the two words was presented in the context of the study list. Figure 2 presents the major findings. The increase in theta band power was significantly larger during the encoding of those words that were later remembered as compared to those which were remembered later. In addition, during recognition, the increase in theta was larger for recognized targets than distractors or not recognized targets.

In all of these studies, subjects knew that their memory will be tested later. Hence, the finding that theta synchronization is significantly stronger during the encoding of those words that can later be remembered could be due to specific encoding strategies. If this were the case, theta synchronization during encoding would reflect a rather unspecific factor and not the actual encoding of new information. One way to avoid the possible influence of encoding strategies is to use an incidental instead of an intentional memory paradigm. The usual procedure is that during the encoding phase of an incidental memory paradigm, subjects do not know that memory performance will be tested later, because they are performing some type of distractor task. Therefore, specific mnemonic techniques and specific attentional factors cannot play a significant role during encoding. Klimesch et al. (1996) used an incidental memory paradigm that consisted of two parts. In the first part, subjects were asked to categorize a series of words and to respond with 'yes' if a word belonged to the category 'living' and with 'no' if a word belonged to the category 'nonliving'. At this point of the experiment, subjects did not know that in the second part of the experiment they would have to recall the words that were presented in the judgment task. Band power values during the encoding stage were calculated and words that could be remembered later were compared with those words that could not be remembered later. The results again indicated that an event-related increase in theta reflects the actual encoding of new information, even with incidental learning of the stimulus words.

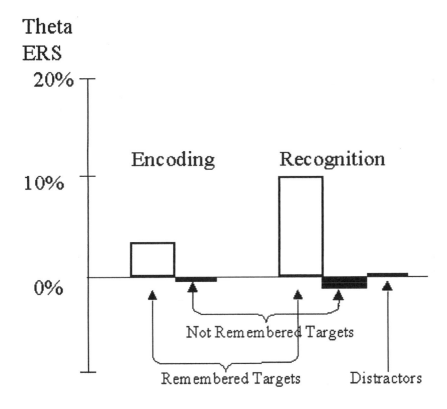

Figure 2. Memory related increases in theta band power during encoding and recognition of words. During encoding of those words that are later remembered in a recognition task, the increase in theta band power is significantly larger than for those that cannot later be remembered. The increase in theta during recognition is larger for words that can be remembered compared to not remembered or new words (after Klimesch et al., 1997c).

The reported findings suggest that theta is a promising neural correlate of episodic and working memory in general (Sarnthein et al., 1998). Although the neural generators of this EEG frequency are not yet known, these findings are in accord with the hypothesis that episodic memory processes are reflected by theta oscillations in complex hippocampo-cortical reentrant loops (Miller, 1991). This assertion is also consonant with converging evidence for the existence of human theta oscillations that comes from studies of patients using depth (Arnolds et al., 1980) and subdural (Kahana et al., 1999) electrodes, as well as from studies with normal subjects using high resolution EEG (Gevins et al., 1997) and MEG (Tesche & Karhu, 2000).

3.2   DELTA ACTIVITY IN THE HUMAN SCALP EEG AND MEMORY PROCESSES

Although it has been found repeatedly that the extent of a memory related increase in band power is largest for the theta band, there is evidence that the delta band shows similar although attenuated results. As an example, recent findings have demonstrated that an increase in theta band power is not restricted to verbal material and is particularly strong during retrieval as compared to encoding (Klimesch et al., 2001). The experiment consisted of two parts, an encoding and a retrieval session. In the first part, a set of 120 pictures was presented one at a time and subjects were instructed to remember them. In the second part, the 'old' pictures were presented randomly intermixed with 120 'new' pictures. Subjects had to decide whether a particular picture was presented previously during the study session. Figure 3 illustrates the results: a large increase in theta band power (within a frequency band of about 4-6 Hz) was found during encoding and retrieval. The largest increase (about 150%) was found during retrieval and for those pictures that were correctly retrieved from memory. Most interesting, however, is the finding that the delta band (about 2-4 Hz) shows similar results, whereas the lower alpha-1 band reflects largely diminished effects with a weak tendency of desynchronization during the late poststimulus interval.

4.   MEMORY CONSOLIDATION IN SLEEP, DELTA AND THETA ACTIVITY

An enduring, related hypothesis is that sleep helps to consolidate memories (Jenkins & Dallenbach, 1924). However, that memory consolidation during sleep is the result of an interaction between the type of memory and sleep stage has only recently been reported (for a review, see Born & Plihal, 2000). As an example, there is increasing evidence that explicit or episodic memory is consolidated during slow wave sleep (SWS or stage 4 sleep) in early sleep, whereas implicit (or procedural) memory is consolidated during REM sleep, which dominates in late sleep. Most interesting, in early as well as in late sleep hippocampal activity appears to play a crucial role for consolidation but in different ways. Traditional research on memory consolidation has focused on REM sleep, with the basic idea that REM deprivation is the major cause for lowered memory performance. The procedure was to wake subjects during the experimental night when the EEG indicated the beginning of REM and to compare memory performance with a control group that experienced undisturbed sleep. Although REM deprivation may very well have unspecific detrimental effects on memory—which most likely are due to increased irritation, mood changes or decreased attention (cf. Van Hulzen,

1986; Van Hulzen & Coenen; Cipolli, 1995), the experimental findings with respect to episodic memory consolidation in human subjects were rather inconclusive. Whereas some studies found lowered episodic memory performance after REM deprivation (Cartwright, 1972; Karni et al., 1994; Lewin & Glaubman, 1975; Tilley & Empson, 1978), others did not find any effects (Castaldo et al., 1974; Ekstrand et al., 1971; Feldman & Dement, 1968; Greenberg et al., 1983; Muzio et al., 1972; Tilley & Empson, 1981).

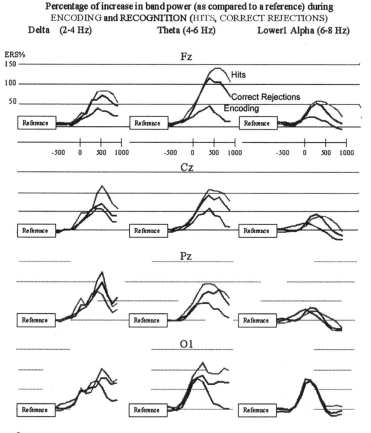

Figure 3. Memory related increases in band power during encoding and recognition of pictures. On the average, the largest increase in band power and the largest differences between conditions (encoding, hits and correct rejections) can be observed in the theta band. However, particularly at Pz, the delta band also shows large effects that are similar to the theta band. This outcome is in sharp contrast to the lower alpha-1 band that shows only a small event-related increase in band power and small differences between conditions. Each band has a width of 2 Hz. Frequency limits represent group averages adjusted for individual alpha frequency from each subject (after Klimesch et al., 2001).

For these reasons, REM deprivation is not an adequate method to study memory consolidation during sleep, and more consistent findings have been obtained when the effects of early and late sleep (first versus second half of the experimental night) were distinguished (Ekstrand et al., 1977). This outcome is important because SWS dominates in early sleep (up to 40% of the first 3 or 4 hours of sleep show delta activity), whereas REM dominates during late sleep. Of particular interest are recent studies performed by Born, Plihal, and colleagues (Born & Fehm, 1998; Born et al. 1991; 1999; Plihal & Born, 1997; Plihal & Born, 1999a; Plihal & Born, 1999b; Plihal et al., 1999). For example, Plihal and Born (1997) orthogonally combined two experimental conditions, early vs. late sleep and episodic vs. implicit memory demands (word-pair association vs. mirror drawing). In the 'early' condition, subjects had to learn a memory task before sleep onset and were awakened after about 3 hours for testing (retrieval). In the 'late' condition subjects were awakened about 3 hours after sleep onset to learn a task and were awakened about 3 hours later for testing. The same procedure was used for control groups with the exception that subjects were not allowed to sleep during the time when early or late sleep occurred. A clear pattern of results was obtained: Increase in episodic memory performance was about twice as much in the early compared to the late condition and the respective control group. In contrast, the increase in implicit memory performance was about twice as much in the late compared to the early condition and the control group. These findings have been replicated by using different episodic (nonverbal instead of verbal memory) memory tasks (Plihal & Born, 1999a).

4.1 THE INVOLVEMENT OF THE HIPPOCAMPAL FORMATION: HIPPOCAMPAL REPLAY AND DELTA ACTIVITY

Given these findings, it appears plausible to assume that memory consolidation during early and late sleep is associated with specific functions of the hippocampal formation and related brain structures. First, during early SWS cortisol secretion reaches its minimum during the circadian cycle. Second, the hippocampal formation and related regions of the temporal lobe have a comparable high density of receptors for corticosteroids. Third, increased levels of cortisol are known to inhibit hippocampal functions via glucocorticoid receptors (Joels & DeKloet, 1994), to reduce neurogenesis of granule cells (Gould et al., 1998), and to reduce memory performance (Pavlides et al., 1995). Moreover, for Plihal and Born (1999b) were able to demonstrate that an experimental increase of cortisol (by infusion) during early sleep reduces episodic memory performance but has no effects during late sleep with implicit memory performance unaffected. In addition, it was shown that these detrimental effects are mediated by glucocorticoid but not by mineralocorticoid receptors.

An important question remains as to how the hippocampal formation is involved in memory consolidation during SWS and REM sleep. Buzsaki (1989, 1996, 1998) has suggested that the establishment of highly selective excitatory network connections (e.g., by LTP or LTD) in the CA3 collateral matrix is that area where episodic information is transiently stored (Wilson & McNaughton, 1994). In this context, it is important to note that after periods of pronounced theta, large irregular field potentials (or SPWs) can be observed (see Figure 1), which have a frequency characteristic in the delta range. It is assumed that SPWs are generated after theta ceases in that part of the hippocampal formation where highly selective excitatory network connections (reflecting freshly encoded information) were established. It is for this reason that SPWs might be considered correlates of memory consolidation.

This hypothesis is supported by several lines of evidence. For example, Wilson and McNaughton (1994) recorded spike trains from 50 to 100 single cells in area C1 of the rat hippocampus during learning of a food reinforced spatial memory task, as well as during SWS sleep before and after task performance. According to the classical work of O'Keefe and Dostrovsky (1971), the preferred location for a given cell to fire ('place field') was determined. In a next step, cells with overlapping place fields were selected and cross-correlations computed. The results show that during encoding (performance of the memory task) cells with overlapping fields exhibit a highly correlated and rhythmic firing pattern within the theta frequency range. The crucial finding, however, is that in SWS sleep after task performance the activity of these cells was also highly correlated, but now the firing pattern was no longer rhythmic but irregular with a frequency in the delta range. In SWS sleep before task performance, these cells did not show correlated activity. In contrast, cells with non-overlapping fields were neither exhibiting rhythmic bursting during encoding nor correlated activity in the delta frequency range during SWS sleep after the memory task. Thus, hippocampal delta activity appears directly related to memory processing of new information.

5.   CONCLUDING REMARKS

The reported findings from animal research and the human scalp EEG clearly support the hypothesis that theta synchronization is related to the encoding and retrieval of new information. Convergent evidence suggests that theta is related to episodic but not semantic memory. One reason for this conclusion is that only theta responds to episodic memory demands, whereas upper alpha responds to semantic memory demands. Furthermore, it is well established that the hippocampal formation together with a complex network involving neocortical and limbic areas are crucial regions for episodic memory processes (Markowitsch, 1996). Tulving (1984) has theorized that episodic memory stores that type of contextual information that keeps an individual autobiographically oriented within space and time. Because time changes the autobiographical context permanently, there is a permanent and vital need to update and store episodic information. Thus, the formation of episodic memory traces is closely related to periods of increased conscious awareness and increased working memory demands.

If we consider the model of hippocampal activity outlined in Figure 1 (Buzsaki, 1996), two states must be distinguished. One state is defined by increased theta activity and is related to increases in episodic memory demands, which dominate in periods of increased conscious awareness (or exploratory behavior in animals). The other state is defined by SPWs or delta activity. It appears to reflect memory consolidation and is related to decreased episodic memory demands and decreased states of conscious awareness. Taken together it can be concluded that delta activity during SWS sleep—which is a state of decreased conscious awareness—reflects consolidation of episodic memory. On the other hand, increased theta activity during REM sleep—which is a state of increased conscious awareness (although of a different type)—provides the necessary contextual 'embedding' for the encoding of implicit memory. It is therefore likely that the lack of context during SWS is helpful for the consolidation of episodic memory, whereas the establishment of a (dream) context during REM is helpful for the consolidation of implicit memory.

Finally, some apparently contradictory results should be discussed. Many studies have reported that absolute (tonic) power in the theta frequency range increases with age (Cristian, 1984; Niedermeyer, 1993b) and is increased in demented subjects (Brenner et al., 1986; Coben et al., 1985). Because memory performance decreases with age (and is decreased in demented subjects), these findings seem to contradict the hypothesis that an event-related increase in theta power reflects episodic memory performance. However, it was demonstrated that event-related theta must be distinguished from tonic power as measured during rest or a reference interval preceding task performance. Indeed, tonic and event-related (or phasic) theta power behave in different ways (Klimesch, 1999). Doppelmayr et al. (1998b) demonstrated, that subjects with large tonic theta power show a significantly smaller event related increase than subjects with smaller tonic power who show a large event-related increase. Thus, tonic and phasic theta power are negatively correlated. Consequently, the finding of an age related increase in tonic theta power is quite consistent with the reported results about a memory related increase in (phasic) theta power.

REFERENCES

Arnolds, D.E.A.T., Lopes Da Silva, F.H., Aitinik, J.W., & Boeijinga, W. (1980). The spectral properties of hippocampal EEG related to behavior in man. *Electroencephalography and Clinical Neurophysiology, 50*, 324-328.

Bland, B.H. (1986). The physiology and pharmacology of hippocampal formation theta rhythms. *Progress in Neurobiology, 26*, 1-54.

Born, J., & Fehm, H.L. (1998). Hypothalamus-pituitary-adrenal activity during human sleep: a coordinating role for the limbic hippocampal system. *Experimental and Clinical Endocrinology & Diabetes, 106*, 153 - 164.

Born, J., DeKloet, R., Wenz, H., Kern, W., & Fehm, H.L. (1991). Changes in slow wave sleep after glucocorticoids and antimineralocorticoids: A cue for central type I corticotropin releasing hormone in humans. *Journal of Clinical Endocrinology and Metabolism, 23*, 126-130.

Born, J., Hansen, K., Marshall, L., Moölle, M., & Fehm, H.L. (1999). Timing the end of nocturnal sleep. *Nature, 397*, 29 - 30.

Born, J., & Plihal, W. (2000). Gedächtnisbildung im Schlaf: Die Bedeutung von Schlafstadien und Streßhormonfreisetzung. *Psychologische Rundschau, 51*, 198-208.

Brenner, R.P., Ulrich, R.F., Spiker, D.G., Scalbassi, R.J., Reynolds, C.F., III, Marin, R.S., & Boller, F. (1986). Computerized EEG spectral analysis in elderly normal, demented and depressed subjects. *Electroencephalography and Clinical Neurophysiology, 64*, 483-492.

Buzsaki, G., Leung, L., & Vanderwolf, C. (1983). Cellular bases of hippocampal EEG in the behaving rat. *Brain Research Reviews, 6*, 139-171.

Buzsaki, G., Bragin, A., Chrobak, J.J., Nadasdy, Z., Sik, A., Hsu, M., & Ylinen, A. (1994) Oscillatory and intermittent synchrony in the hippocampus: Relevance to memory trace formation. In G. Buzsaki, R. Llinas, W. Singer, A. Berthoz, & Y. Christen (Eds.), Temporal coding in the brain (pp. 145-172). Berlin, Heidelberg: Springer-Verlag.

Buzsaki, G. (1989). A two-stage model of memory trace formation: A role for "noisy", brain states. *Neuroscience, 31*, 551 - 570.

Buzsaki, G. (1996). The hippocampo-neocortical dialogue. *Cerebral Cortex, 6*, 81-92.

Buzsaki, G. (1998). Memory consolidation during sleep: A neurophysiological perspective. *Journal of Sleep Research, 7*, 17-23.

Castalado, V., Krynicki, V., & Goldstein, J. (1974). Sleep stages and verbal memory. *Perceptual and Motor Skills, 39*, 1023 - 1030.

Cartwright, R.D. (1972). Problem solving in REM, NREM, and waking. *Psychophysiology, 9*, 108.

Christian, W. (1984). Das Elektroencephalogramm (EEG) im höheren Lebensalter. *Nervenarzt, 55*, 517-524.

Coben, L.A., Danziger, W., & Storandt, M. (1985). A longitudinal EEG study of mild senile dementia of Alzheimer type: Changes at 1 year and at 2.5 years. *Electroencephalograph and Clinical Neurophysiology, 61*, 101-112.

Doppelmayr, M., Klimesch, W., Pachinger, Th., & Ripper, B. (1998a). Individual differences in brain dynamics: Important implications for the calculation of event-related band power measures. *Biological Cybernetics, 79*, 49-57.

Doppelmayr, M., Klimesch, W., Pachinger, Th., & Ripper, B. (1998b). The functional significance of absolute power with respect to event-related desynchronization. *Brain Topography, 11*, 133-140.

Eckstrand, B.R., Sullivan, M.G., Parker, D.F., & West, J.N. (1971). Spontaneous recovery and sleep. *Journal of Experimental Psychology, 88*, 142 - 144.

Eckstrand, B.R., Barett, T.R., West, J.N., & Maier, W.G. (1977). The effect of sleep on human long-term memory. In R.R. Drucker-Colin & J.L. McGaugh (Eds.), *Neurobiology of sleep and memory* (pp. 419 - 438). New York: Academic Press.

Feldman, R., & Dement, W. (1968). Possible relationships between REM sleep and memory consolidation. *Psychophysiology, 5*, 243 - 251.

Gevins, A., Smith, M.E., McEvoy, L., & Yu, D. (1997). High-resolution EEG mapping of cortical activation related to working memory: effects of task difficulty, type of processing, and practice. *Cerebral Cortex, 7*, 374-385.

Givens, B. (1996). Stimulus-evoked resetting of the dentate theta rhythm: relation to working memory. *NeuroReport, 8*, 159-163.

Gould, E., Tanapat, P., McEwen, B., Flügge, G., & Fuchs, E. (1998). Proliferation of granule cell precursors in the dentate gyrus of adult monkeys is diminished by stress. *Proceedings of the National Academy of Science of the USA, 95*, 3168-3171.

Greenberg, R., Pearlman, C., Schwartz, W.R., & Grossman, H. (1983). Memory, emotion, and REM sleep. *Journal of Abnormal Psychology, 92*, 378-381.

Greenstein, Y.J., Pavlides, C., & Winson, J. (1988) Long-term potentiation in the dentate gyrus is preferentially induced at theta rhythm periodicity. *Brain Research, 438*, 331-334.

Jenkins, J.G., & Dallenbach, K.M. (1924). Oblivicence during sleep and waking. *American Journal of Psychology, 35*, 605 - 612.

Kahana, M.J., Sekuler, R., Caplan, J.B., Kirschen, M., & Madsen, J.R. (1999). Human theta oscillations exhibit task dependence during virtual maze navigation. *Nature, 399*, 781-784.

Karni, A., Tanne, D., Rubenstein, B.S., Askenasy, J.J.M., & Sagi, D. (1994). Dependence on REM sleep of overnight improvement of a perceptual skill. *Science, 265*, 679 - 681.

Klimesch, W. (1994). *The structure of long-term memory: A connectivity model for semantic processing*. Hillsdale, NJ: Lawrence Erlbaum.

Klimesch, W. (1995). Memory processes described as brain oscillations in the EEG-alpha and theta band. *Psycoloquy.*95.6.06. (electronic journal) memory-brain.1.klimesch.

Klimesch, W. (1996). Memory processes, brain oscillations and EEG synchronization. *International Journal of Psychophysiology, 24*, 61-100.

Klimesch, W., Schimke, H., & Schwaiger, J. (1994). Episodic and semantic memory: An analysis in the EEG-theta and alpha band. *Electroencephalography and Clinical Neurophysiology, 91*, 428-441.

Klimesch, W., Doppelmayr, M., Russegger, H., & Pachinger, Th. (1996). Theta band power in the human scalp EEG and the encoding of new information. *NeuroReport, 7*, 1235 - 1240.

Klimesch, W., Doppelmayr, M., Pachinger, Th., & Ripper, B. (1997a). Brain oscillations and human memory performance: EEG correlates in the upper alpha and theta bands. *Neuroscience Letters, 238*, 9-12.

Klimesch, W., Doppelmayr, M., Pachinger, Th., & Russegger, H. (1997b). Event-related desynchronization in the alpha band and the processing of semantic information. *Cognitive Brain Research, 6*, 83-94.

Klimesch, W., Doppelmayr, M., Schimke, H., & Ripper, B. (1997c). Theta synchronization in a memory task. *Psychophysiology, 34*, 169-176.

Klimesch, W., Vogt, F., & Doppelmayr, M. (2000). Interindividual differences in alpha and theta power reflect memory performance. *Intelligence, 27*, 347-362.

Klimesch, W., Doppelmayr, M., Stadler, W., Pöllhuber, D., & Röhm, D. (2001). Episodic retrieval is reflected by a process specific increase in human EEG theta activity. *Neuroscience Letters, 302*, 49-52.

Kroll, N.E.A., & Klimesch, W. (1992). Semantic memory. Complexity or connectivity? *Memory & Cognition, 20*, 192-210.

Larson, J., Wong, D., & Lynch, G. (1986). Patterned stimulation at the theta frequency is optimal for the induction of hippocampal long-term potentiation. *Brain Research, 368*, 347-350.

Lisman, J.E., & Idiart, M.A.P. (1995). Storage of 7±2 short-term memories in oscillatory subcycles. *Science, 267*, 1512-1515.

Lopes da Silva, F.H. (1992) The rhythmic slow activity (theta) of the limbic cortex: An oscillation in search of a function. In E. Basar & T.H. Bullock (Eds.), *Induced rhythms in the brain* (pp. 83-102). Boston: Birkhäuser.

Lewin, I., & Glaubman, H. (1975). The effect of REM deprivation: Is it detrimental, beneficial, or neutral? *Psychophysiology, 12*, 349 - 353.

Maren, St., DeCola, J., Swain, R., Fanselow, M., & Thompson, R. (1994). Parallel Augmentation of hippocampal long-term potentiation, theta rhythm, and contextual fear conditioning in water-deprived rats. *Behavioral Neuroscience, 108*, 44-56.

Markowitsch, H. (1996). Neuropsychologie des Gedächtnisses. *Spektrum der Wissenschaft, 9*, 52-61.

Miller, R. (1991). Cortico-hippocampal interplay and the representation of contexts in the brain. Berlin: Springer.

Montoya, Ch., Heynen, A., Faris, P., & Sainsbury, R. (1989). Modality specific type 2 theta production in the immobile rat. *Behavioral Neuroscience, 103*, 106-111.

Muzio, J.W., Roffwarg, H.P., Anders, C.B., & Muzio, L.G. (1972). Retention of rote learned meaningful verbal material and alternation in the normal EEG pattern. *Psychophysiology, 9*, 108.

Niedermeyer, E. (1993). Normal aging and transient cognitive disorders in the elderly. In E. Niedermeyer & F.H. Lopes da Silva (Eds.), *Electroencephalography: Basic principles, clinical applications, and related fields*, (pp. 329-338). Baltimore: Williams & Wilkins.

O'Keefe, J. (1993). Hippocampus, theta and spatial memory. *Current Opinion in Neurobiology, 3*, 917-924.

O'Keefe, J., & Dostrovsky, J. (1971). The hippocampus as a spatial map: Preliminary evidence from unit activity in freely moving rats. *Brain Research, 34*, 171-175.

O'Keefe, J., & Recce, M. (1993). Phase relationship between hippocampal place units and the EEG theta rhythm. *Hippocampus, 3*, 317-330.

O'Keefe, J., & N. Burgess, N. (1999). Theta activity, virtual navigation and the human hippocampus. *Trends in Cognitive Science, 3*, 403-406.

Pavlides, C., Watanabe, Y., Magarinos, A.M., & McEwen, B.S. (1995). Opposing roles of type I and Type II adrenal steroid receptors in hippocampal long-term potentiation. *Neuroscience, 68*, 387-394.

Petsche, H., Stumpf, C., & Gogolak, G. (1962). The significance of the rabbit's septum as a relay station between the midbrain and the hippocampus. *Electroencephalography and Clinical Neurophysiology, 19*, 25-33.

Pfurtscheller, G., & Aranibar, A. (1977). Event-related cortical synchronization detected by power measurements of scalp EEG. *Electroencephalography and Clinical Neurophysiology, 42*, 817-826.

Plihal, W., & Born, J. (1997). Effects of early and late nocturnal sleep on declarative and procedural memory. *Journal of Cognitive Neuroscience, 9*, 534 -547.

Plihal, W., & Born, J. (1999a). Effects of early and late nocturnal sleep on priming and spatial memory. *Psychophysiology, 36*, 571-582.

Plihal, W., & Born, J. (1999b). Memory consolidation in human sleep depends on inhibition of glucocorticoid release. *Neuroreport, 10*, 2741-2747.

Plihal, W., Pietrowsky, R., & Born, J. (1999). Dexamethasone blocks sleep induced improvement of declarative memory. *Psychoneuroendocrionology, 24*, 312-331.

Sarnthein, J., Petsche, H., Rappelsberger, P., Shaw, G.L., & von Stein, A. (1998). Synchronization between prefrontal and posterior association cortex during human working memory. *Proceedings of the National Academy of Science of the USA, 95*, 7092-7096.

Schacter, D. (1977). EEG theta waves and psychological phenomena: A review and analysis. *Biological Psychology, 5*, 47-82.

Scoville, W., & Milner, B. (1957). Loss of recent memory after bilateral hippocampal lesions. *Journal of Neurology, Neurosurgery and Psychiatry, 20*, 11-21.

Skaggs,W.E., & McNaughton, B.L. (1996). Replay of neuronal firing sequences in rat hippocampus during sleep following spatial experience. *Science, 271*, 1870 - 1873.

Squire, L.R. (1992). Memory and the hippocampus: A synthesis from findings with rats, monkeys, and humans. *Psychological Review, 99*, 195-231.

Tesche, C.D., & Karhu, J. (2000). Theta oscillations index human hippocampal activation during a working memory task. *Proceedings of the National Academy of Science of the USA, 97*, 919-924.

Tilley, A.J., & Empson, J.A. (1978). REM sleep and memory consolidation. *Biological Psychology, 6*, 293-300.

Tilley, A.J. & Empson, J.A. (1981). Picture recall and recognition following total and selective sleep deprivation. In W.P. Koella (Ed.), *Sleep '80* (pp. 367-369). Basel: Karger.

Tulving, E. (1984). Precis of elements of episodic memory. *Behavioral and Brain Sciences, 7*, 223-268.

Vanderwolf, C.H. (1992). The electrocorticogram in relation to physiology and behavior: A new analysis. *Electroencephalography and clinical Neurophysiology, 82*, 165-175.

Vanderwolf, C., & Robinson, T. (1981). Retico-cortical activity and behavior: A critique of the arousal theory and a new synthesis. *Behavioral and Brain Sciences, 4*, 459-514.

VanHulzen, Z.J.M. (1986). *Paradoxical sleep deprivation and information processing in the rat.* Thèse, Université de Nujmegen, Nijmegen.

VanHulzen, Z.J.M., & Coenen, A.M. (1980). The pendulum technique for PS deprivation in rats. *Physiology and Behavior, 25*, 807-811.

Wilson, M.A., & McNaughton, B.L. (1994). Reactivation of hippocampal ensemble memories during sleep. *Science, 265*, 676-679.

Winson, J. (1990). The meaning of dreams. *Scientific American, 263*, 42-48.

Chapter 10

GAMMA ACTIVITY IN THE HUMAN EEG

CHRISTOPH S. HERRMANN
Max-Planck-Institute of Cognitive Neuroscience, Leipzig

Detection of change is one of the most prominent obligations of the human brain and can occur in any sensory modality. For example, when a young child is shown the same red card repeatedly, the child will lose interest in it—i.e., the response habituates. However, as a blue card is presented, the new stimulus will immediately result in responses signaling increased attention (Squire & Kandel, 1999). This pattern of habituation/attention is adaptive, since the processing system should not react to perceptions that stay constant over time. Instead, the system should focus on events that change suddenly and indicate the need for a response. This process reflects automatic or bottom-up attention. In addition, selective or top-down attention can be engaged such that a voluntary focus on selected portions of auditory and visual input occurs. The present chapter will review how high-frequency electroencephalographic (EEG) activity or "gamma" is germane to the attentional mechanisms underlying the detection of change.

1. OSCILLATIONS IN THE EEG

EEG analysis is one of the main methods used to investigate the functional behavior of the human and animal brain. Although physicians focus on continuous and relatively gross EEG recordings, specific sensory and cognitive processes can be measured by averaging EEG responses to stimuli to extract event-related potentials (ERPs). Both EEG and ERP measures can be investigated in the frequency domain, and it has been convincingly demonstrated that assessing specific frequencies can often yield insights into the functional cognitive correlations of these signals (Başar et al., 1999). This result can be achieved by selectively filtering out those parts of the signal that oscillate at a given frequency. Since, in principle, every signal can be decomposed into sinusoidal oscillations of different amplitudes; such decomposition is usually computed using the Fourier transform to quantify the oscillations that constitute the signal (Dumermuth, 1977).

1.1   ALPHA RHYTHM BEGINNINGS

Oscillations were the very beginning of EEG research when the German neurophysiologist Berger (1929) first observed the dominant oscillations of approximately 10 Hz recorded from the human scalp. Berger coined the term alpha frequency for activity in this frequency range by using the first letter of the Greek alphabet.

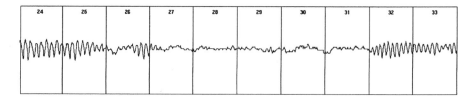

Figure 1. Ten seconds of continuous EEG recorded with eyes closed showing slow alpha activity (7-8 Hz) in seconds 24 and 25. Eye-opening in seconds 26 to 31 results in suppression of alpha activity and subsequently leads to speeded alpha of 10-11 Hz in seconds 32 and 33.

Berger dubbed the second type of rhythmic activity that he found in the human EEG as beta, which is now considered to be the frequency range of approximately 12-30 Hz. Following this consecutive ordering, Adrian (1942) referred to the oscillations around 40 Hz observed after odor stimulation in the hedgehog as gamma waves. Başar-Eroglu et al. (1996b) describe this report as the first stage of gamma research. In this taxonomy, the second stage was initiated by Freeman (1975), who found 40 Hz was strongly associated with perceptual models of the rabbit's olfactory bulb. The third phase started with the work of Galambos et al. (1981), which made gamma oscillations generally accepted in studies of human perception. The fourth phase and a major influence of gamma activity research stemmed from Gray et al. (1989), who showed that synchronous firing of single neurons in the 40 Hz range could help account for the 'perceptual binding' that produces a unitary conscious experience. Karakaş and Başar (1998) have helped to define the fifth phase, which is marked by an enormous number of different paradigms and methods applied to solve the 'gamma puzzle'.

1.2 GAMMA ACTIVITY AND ITS FUNCTIONAL ROLES

Neurons in primary sensory cortex typically code simple features of perceived stimuli, such that perceptual objects are composed of various features, which are represented by different neurons in the brain. The neural activity that codes objects features is somehow bound together, to produce the perception and a coherent object. The so-called binding problem arises when multiple objects are perceived at one time, and their single features could potentially be bound incorrectly to produce illusory conjunctions. Thus, understanding the neurocognitive mechanisms of binding and attention is a fundamental and important cognitive problem.

Oscillatory activity in the gamma frequency range (30-80 Hz) has been found to reveal correlates of processes that are associated with binding phenomena. In particular, neurons in the animal brain that oscillate at about 40 Hz are believed to represent the binding of different features of one object to form a single coherent percept (Eckhorn et al., 1988; Engel et al., 1992; Gray et al., 1989). Figure 2 schematically illustrates this process. When one bar (object) is moved across the receptive fields of two neurons in cat visual cortex, the responses of these two neurons are synchronous (i.e., they spike at the same time) and their frequency occurs in the gamma range. When two bars move in the same direction (and are usually perceived as one interrupted object) the neurons still fire with some degree of synchrony. However, if two bars move in opposite directions, which will be perceived as two individual objects, the neural discharges are no longer synchronous.

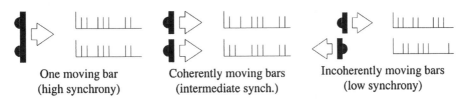

| One moving bar | Coherently moving bars | Incoherently moving bars |
| (high synchrony) | (intermediate synch.) | (low synchrony) |

Figure 2. Black bars moving across the receptive fields (gray) of neurons in cat visual cortex and the neuronal response. Vertical lines indicate single-unit spikes in response to stimuli.

Similar findings have been reported from the human EEG, which demonstrate higher induced gamma activity for one bar than for two incoherently moving bars (Müller et al., 1997). Such findings have been interpreted as reflecting gamma activity in the human EEG that is associated with visual binding. It has been found for illusory contours (Kanizsa figures, Section 4) where the inducer disks are bound together for the perception of the figure (Hermann et al., 1999; Tallon et al., 1995; Tallon-Baudry et al., 1996). Induced gamma activity also has been reported when subjects suddenly see a meaningful object in formerly non-meaningful stimulus (Keil et al., 1999; Tallon-Baudry et al., 1997).

Another function reflected by gamma activity is attention. Tiitinen et al. (1993) demonstrated that the gamma response around 50 milliseconds after auditory tone pips was larger when subjects were instructed to attend to the ear where the stimulus occurred (attended condition) compared to when they were instructed to attend to the other ear (unattended condition). Similar findings in the visual modality revealed increased gamma activity over the occipital cortex for attended versus unattended flickering lights (Müller et al., 1998). Data from experiments with different Kanizsa figures designed to differentiate binding and attention processes also support the notion that attention is a main source for gamma activity (Hermann & Mecklinger, 2000a; Hermann & Mecklinger, 2000b; Hermann et al., 1999).

1.3   FURTHER ROLES OF GAMMA ACTIVITY

In addition to binding and attention as functional correlates of gamma, other processes also have been associated with gamma activity. Jokeit and Makeig (1994) and Müller et al. (1998) have shown how gamma activity correlates with reaction times of fast and slow responders. Başar-Eroglu et al. (1996a) related gamma activity with the formation of a stable percept in a multi-stable pattern (a Necker cube) by showing that before a pattern reversal, there is more gamma activity than during the stable perception of one of the two percepts. Miltner et al. (1999) have demonstrated how associative learning produces coherence of gamma activity in visual and motor areas when motor responses to light stimuli are learned. Reviews related to the functional relevance of gamma oscillations in humans and animals can be found in Başar-Eroglu et al. (1996b). Additional reviews concerning the relation of gamma activity to human visual perception (Tallon-Baudry & Bertrand, 1999) and attentional mechanisms (Müller et al., 2000) are also available.

2.   TYPES OF GAMMA ACTIVITY

According to a classification of different types of gamma activity by Galambos (1992), there are spontaneous, induced, and evoked gamma rhythms, all of which are differentiated by their degree of phase-locking to the stimulus (emitted gamma rhythms in response to omitted stimuli also have been observed, but these will not be considered here). In this framework, spontaneous activity is completely uncorrelated with the occurrence of an experimental condition. Induced activity is correlated with experimental conditions but is not strictly phase-locked to its onset. Evoked activity is strictly phase-locked to the onset of an experimental condition.

Figure 3 illustrates phase-locking. It is important to note that phase-locking, rather than time-locking, is the crucial parameter that determines whether or not activity is cancelled out or summed when multiple signals are analyzed. Averaging signals with temporal relations, as illustrated in Figure 3a and b or c and d, they will sum since they are effectively phased-locked to the virtual stimulus at time point 125. Alternatively, if signals are time-locked but not phase-locked, as in Figure 3a and c, or neither time- or phase-locked, as in Figure 3b and d, single trials will cancel out in the average.

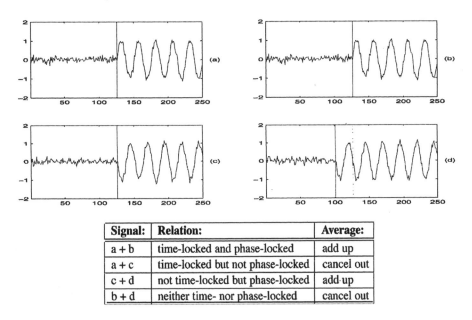

Signal:	Relation:	Average:
a + b	time-locked and phase-locked	add up
a + c	time-locked but not phase-locked	cancel out
c + d	not time-locked but phase-locked	add up
b + d	neither time- nor phase-locked	cancel out

Figure 3. Signals must be phase-locked, not time-locked to sum across multiple epochs.

Some spurious oscillations in the gamma frequency range are present in the human EEG without correlation to experimental conditions during and between stimulation periods. This activity is considered to be spontaneous and usually cancels out completely if an averaged ERP is computed across enough stimulus repetitions. Thus, true oscillations likely originate from cognitive processes unrelated to the specific mental task being performed.

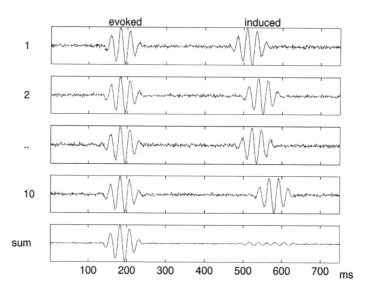

Figure 4. If oscillations occur at the same latency after stimulus onset with the same phase relative to stimulus onset in multiple trials (rows 1-4), they are considered evoked by the stimulus (left). If latency or phase jitter relative to stimulus onset, the oscillations are considered to be induced by the stimulus (right). Evoked activity sums up in the average (bottom row), while induced activity is nearly cancelled out.

If oscillations occur after each stimulation but with varying onset times and/or phase jitter, they are considered as being induced by the stimulus rather than evoked and are not visible in the averaged ERP. Figure 3 (right) illustrates this outcome. Special methods have to be applied to record this type of activity (see Section 3). This type of gamma activity is assumed to reflect cognitive processes of binding and figure representation.

Oscillatory activity in EEG can be phase-locked to the onset of an experimental stimulus, as it starts at approximately the same latency after stimulus onset for every repetition of the stimulus. Figure 3 (left) illustrates this outcome. In this case, the activity is called evoked, sums, and is visible in the averaged ERP. This type of activity is usually evoked by any kind of sensory stimulation, like auditory, visual or somatosensory stimulation.

2.1   LATENCY

Differentiation of gamma activity also can be made according to the latency at which it occurs after stimulus onset. Early gamma activity usually peaks around 100 milliseconds after stimulus onset and is evoked by the stimulus. It reflects early processes of stimulus encoding and attention. Tallon-Baudry et al. (1996, 1997) have shown that gamma activity can peak at a later time. This late gamma activity is usually induced by the stimulus and reflects perceptual and cognitive processes.

3. METHODS FOR GAMMA ANALYSIS

The analysis of gamma and other EEG frequencies requires some precautions when data are recorded as well as specific frequency analysis tools. These are discussed next.

Two important parameters for the recording equipment are critical to properly record gamma activity: (1) The sampling rate has to be set to a value that is at least twice the frequency that should be analyzed (four times is better and is required by some software). For example, if gamma activity up to 80 Hz will be analyzed, a minimum sampling rate of 160 Hz is needed and 320 is recommended. (2) The low pass filter needs to be set to a value higher than the highest frequency that should be analyzed. The low pass filter is usually integrated in the analog amplifier to prevent aliasing errors when digitizing analog data. This step is sometimes overlooked when trying to record gamma activity for the first time. It is also worth noting that the lower 3 dB edge frequency is the critical value of a low-pass filter and not its middle frequency.

3.1 ARTIFACTS

All artifacts that contaminate traditional ERP averages should be excluded from gamma analysis as well. In addition, there are several specific artifact conditions that are especially crucial when gamma activity is analyzed.

A potential problem when recording gamma is the influence of alpha frequency harmonics. Whenever an oscillation is not purely sinusoidal, it leads to so-called harmonic frequencies at integer multiples of that frequency. For non-sinusoidal alpha activity around 10 Hz, one such harmonic can be in the gamma range (Jürgens et al., 1995). Figure 5 (left) illustrates how a pure sine wave of 10 Hz leads to exactly one spectral peak at 10 Hz, but even a slight change of its shape (right) can lead to harmonic peaks at 20 and 40 Hz. Hence, studies should ensure that the gamma activity behaves independently of the alpha activity, and that the alpha activity simply does not show the identical effects. Different latencies and different topographical distributions can serve to discriminate the two outcomes.

In addition, it is important to emphasize that when differences in EEG data are absent between two experimental conditions, this result does not necessarily imply an absence of differences in the underlying brain processes. It may be that the electric responses of the brain processes are different but they do not propagate all the way through brain and skull to affect EEG recordings.

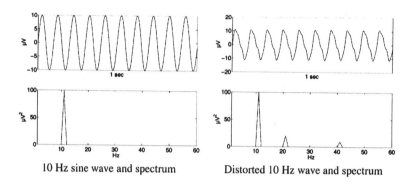

10 Hz sine wave and spectrum   Distorted 10 Hz wave and spectrum

Figure 5. A sinusoidal 10 Hz wave leads to exactly one spectral frequency response at 10 Hz (left). A distorted sine wave can lead to additional harmonic frequencies spectrum (right).

Another potential confound of human gamma activity is electromyography (EMG). If subjects sit uncomfortably or chew during an EEG session and innervate their muscles, the EEG electrodes will record EMG activity. This high frequency muscle-related activity (30-80 Hz) can be mistaken for gamma EEG activity. Therefore, all epochs that are subsequently averaged should be visually evaluated for the occurrence of such EMG artifacts, which should then be excluded from further analysis.

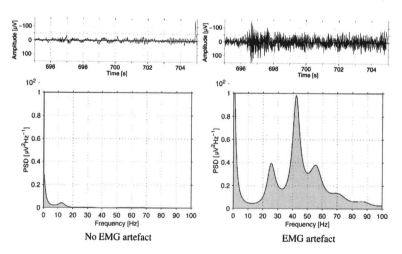

No EMG artefact   EMG artefact

Figure 6. Clean EEG data and its frequency spectrum (left) and an epoch with EMG contamination leading to frequency peaks around 40 Hz.

Figure 6 shows ten seconds of clean EEG and the corresponding frequency spectrum with a 0 Hz and a 12 Hz peak (left). EMG activity can easily be detected in the time domain (right) but may be mistaken for gamma activity in the spectrum.

3.2   FREQUENCY ANALYSIS METHODS

Several methods exist to exclusively extract oscillations of a specific frequency from ERP data. Among the most popular are filtering, Fourier transformation, and wavelet analysis.

Figure 7 shows the results of the three methods to extract frequency information. Left panel: filtering an ERP with a band pass filter (35-45 Hz) shows a clear burst of 40 Hz activity around 100 milliseconds. This oscillatory activity is enhanced for the dotted as compared to the solid condition. Middle panel: Fourier spectrum analyses of the time interval from 50-150 milliseconds. An increase of activity for the dotted condition can be noticed around 40 Hz. Right panel: the absolute values of the wavelet transform of the ERP are shown for a 40 Hz wavelet. The difference between conditions is very prominent and can be observed at every point in time due to the lack of oscillations in the signal. The wavelet transform can be thought of as the envelope of the filtered ERP. The wavelet transform is ideally suited for ERP frequency analysis and will be discussed below.

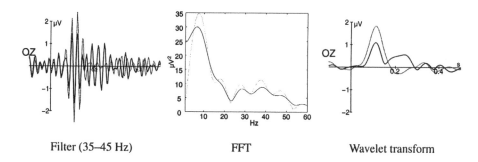

| Filter (35–45 Hz) | FFT | Wavelet transform |

Figure 7. Three possibilities to extract frequency information from ERP data: a 35-45 Hz filtered ERP (left), the FFT spectrum of the epoch 50-150 milliseconds (middle) and the wavelet transform of the ERP (right).

3.3   THE WAVELET TRANSFORM

To compute a wavelet transform, the original signal is convolved with a wavelet function. In the case of the Morlet wavelet used here, it is calculated according to the formula

$$\Psi(t) = e^{j\omega t} \cdot e^{-t^2/2}$$

where ω is 2π times the frequency of the unshifted and uncompressed mother wavelet. Figure 8 schematically illustrates how these mathematical terms construct a wavelet.

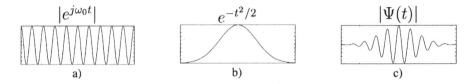

Figure 8. Multiplying a sinusoidal function (a) and an envelope function (b) results in a wavelet (c).

Mathematically convolving wavelets with signals produces a new signal (the convolution) that can be interpreted as the similarity of the wavelet to the signal. Wavelets can be compressed to obtain wavelets of different frequencies (substitute t by t/a, where a = compression factor). The mother wavelet (a = 1) has the same frequency as the sampling frequency (f_s) of the signal. Wavelets of lower frequencies are computed by increasing a (e.g., if $a = f_s$ the wavelet has a frequency of 1 Hz).

Convolving the signal and the shifted and compressed wavelet leads to a new signal

$$s_a(b) = A \int \overline{\Psi}\left(\frac{t-b}{a}\right) \cdot x(t) \ dt$$

where $\overline{\Psi}$ is the conjugate of the complex wavelet and $x(t)$ is the original signal. These new signals $s_a(b)$ are computed for different scaling factors a. For the experiments in Section 4, we calculated the gamma activity by using a wavelet that was compressed to 40 Hz. The scaling factor $A = 1/\sqrt{a}$ is used to scale the wavelet prior to convolution.

To represent phase-locked (evoked) activity, the wavelet transform is computed on the average over the single trials (the ERP). This is denoted by the formula WTAvg (Wavelet Transform of Average). Since the wavelet transform returns complex numbers, the absolute values are calculated.

$$\text{WTAvg} = \left| A \int \overline{\Psi}\left(\frac{t-b}{a}\right) \cdot \frac{1}{n} \sum_{i=1}^{n} eeg_i(t) \ dt \right|$$

The baseline of the raw data in a time interval prior to stimulation needs to be subtracted from each EEG epoch prior to averaging. After calculating the gamma activity, the frequency-specific baseline activity at 40 Hz can be subtracted to yield values that indicate gamma amplitude relative to baseline. When wavelet convolutions are computed, the convolution peaks at the same latency as the respective frequency component in the raw data, although the peak width will be smeared. Therefore, the baseline should be chosen to precede the stimulation by half the width of the wavelet (i.e., 150 milliseconds for six 25 millisecond cycles of a 40 Hz wavelet) to avoid the temporal smearing of post-stimulus activity into the interval directly preceding the stimulus. To avoid distortions by the rectangular window function that can result from 'cutting out' a single epoch from continuous raw data, the convolution should start and end one wavelet length before the baseline and after the end of the assessed time interval.

Figure 9 (left), depicts the convolution of an EEG with a wavelet that results in a new signal. These wavelet convolutions can be computed for multiple frequencies and the amplitudes of the convolutions can then be color- or gray-scale-coded in one single diagram. Figure 9 (right) illustrates this method, which is called a time-frequency representation.

This time-frequency representation (WTAvg) contains only that part of the activity that is phase-locked to the stimulus onset. To compute the activity that is not phase-locked to stimulus onset (and is therefore cancelled out in the average), the sum of evoked and induced activity can be computed. To calculate the sum of all activity at one frequency, the absolute values of the wavelet transforms of the single trials are averaged (AvgWT), which means that each single trial is at first transformed and the absolute values are averaged subsequently.

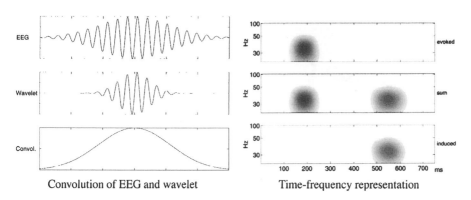

Convolution of EEG and wavelet   Time-frequency representation

Figure 9. Left: Convolving an EEG (top) with a moving wavelet (middle) results in a convolution (bottom). Multiple such convolutions can be mapped in a time frequency representation. This is shown for the evoked gamma activity (top) of the example in Figure 4, the sum of evoked and induced gamma activity (middle) and isolated induced gamma activity (bottom).

This new time-frequency representation contains all activity of one frequency that occurred after stimulus onset, no matter whether it was phase-locked to the stimulus or not. As above, the 40Hz activity in a pre-stimulus interval (e.g., -400 to −150 milliseconds) can be subtracted to obtain a relative measure. This sum of evoked and induced activity is also known simply as induced activity (Tallon-Baudry & Bertrand, 1999), since the absolute amount of evoked activity is small compared to the much higher absolute values of the summed activity. To obtain only activity not phase-locked to stimulus onset, the evoked activity (WTAvg) needs to be subtracted from the sum of evoked and induced activity (AvgWT), such that absolute measures are subtracted to obtain AvgWT-WTAvg (i.e., no baseline correction in the frequency domain).

$$\text{AvgWT} = \frac{1}{n}\sum_{i=1}^{n} \left| \frac{1}{\sqrt{a}} \int \overline{\Psi}\left(\frac{t-b}{a}\right) \cdot x_i(t) \, dt \right|$$

4.   GAMMA AND ATTENTION: AN ILLUSTRATIVE STUDY

To assess whether gamma activity is related to top-down processes of attention, two experiments were conducted using the same stimuli but different tasks. It is assumed that electrophysiological responses that are identical across both experiments reflect bottom-up processes and do not affect top-down task requirements. If electrophysiological responses change between experimental conditions, these must reflect a top-down process.

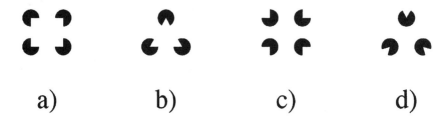

Figure 10. The four stimulus types used in the experiment: a) Kanizsa square, b) Kanizsa triangle, c) non-Kanizsa triangle, and d) non-Kanizsa square.

Figure 10 shows the four stimuli that were used for the two experimental conditions. Two of the stimuli represent Kanizsa figures (a and b), while the others (c and d) have similar physical properties but do not constitute illusory figures. Note that in the latter figures the pac-men are rotated in such a way that none of the separate stimuli can be bound together into shapes by collinear line segments. Therefore, the two experimental conditions are designed to differentiate between binding and attention, since one stimulus is defined as target (to test attention) but two of them require binding for the perception of an illusory figure. In Experiment 1, the Kanizsa square was defined as the target and had to be counted by the subjects (Herrmann et al., 1999). In Experiment 2, the non-Kanizsa square (c) was defined as the target and was counted instead (Herrmann & Mecklinger, 2001).

Figure 11 shows the ERPs to the four stimuli. As expected, the Kanizsa square elicited the largest P3 in Experiment 1, since it was counted. The non-Kanizsa square elicited the largest P3 in Experiment 2. The P3 is the only ERP component that represents a top-down process in these experiments. In Experiment 2, its amplitude was suppressed and latency prolonged, which indicates that the task is harder than in Experiment 1.

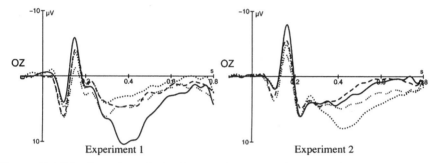

Figure 11. ERPs from electrode Oz for Experiments 1 and 2. P1 and N1 components are independent of task requirements. The P3 component is affected by the task change between experiments and is delayed in latency and reduced in amplitude. Kanizsa square (solid), Kanizsa triangle (dashed), non-Kanizsa square (dotted) and non-Kanizsa triangle (intermittently dotted).

The P1 and N1 components were constant across the two experiments. P1 and N1 reflect the bottom-up processes of sensory input coding. Hence, P1 was mainly affected by the number of pac-men in a figure. Triangles evoke larger P1 amplitudes than the squares, which may be due to less extinction of the unsymmetrical shapes in the two hemispheres. Note that the Kanizsa figures elicited larger N1 amplitudes than the non-Kanizsa figures. This result suggests that the illusory figures are clearly processed by the subjects.

Figure 12 presents the topographic distribution of the early evoked gamma activity (50 to 150 milliseconds) for the four different stimuli in Experiment 1. The Kanizsa square, which was defined as the target stimulus and mentally counted, clearly demonstrated stronger activation than the other three stimuli. This target effect could possibly mean that the early evoked gamma activity reflects a top-down mechanism of attention. Even though the four stimuli varied on two dimensions (figureness and number of pac-men) to differentiate between binding and attention, the use of a Kanizsa figure as a target may have produced a confound, as this stimulus is readily perceived as a "square." However, Experiment 2 employed the non-Kanizsa square as the target, and it was expected that this new target would evoke the early gamma activity as did the Kanizsa square in Experiment 1.

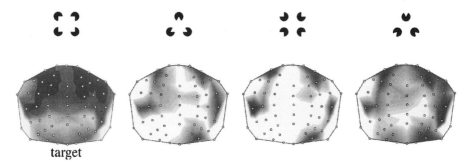

                target

Figure 12. Topographic amplitude maps of the early gamma activity for the four stimuli of Experiment 1. The target stimulus elicited the highest gamma response.

Figure 13 presents the topographic amplitude distribution of the early evoked gamma activity for the stimuli of Experiment 2. The pattern is clearly different—the non-Kanizsa square now elicits the largest gamma response. This result is consistent with the hypothesis that the Kanizsa square and the non-Kanizsa triangle also elicit some gamma activity. These two stimuli share one of the figures of the target stimulus: The Kanizsa square is also a square, and the non-Kanizsa triangle is also a non-Kanizsa figure.

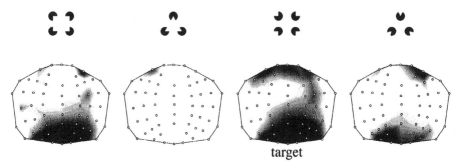

                                          target

Figure 13. Topographic amplitude maps of the early gamma activity for the four stimuli of Experiment 2. Even though the four stimuli are identical to Experiment 1, the change of the task affects the gamma response. The target elicits the largest gamma response.

The results of the early evoked gamma activity demonstrate that just like the P3 ERP, the early evoked gamma activity reflects top-down processes of attention. The P3 peaks around 400 milliseconds, but the early evoked gamma activity peaks much earlier around 100 milliseconds. It is noteworthy that the order of amplitude of the early evoked gamma activity already resembles the pattern of reaction times about 500 milliseconds before the actual button press (Herrmann & Mecklinger, 2001).

5. CONCLUSION

Gamma activity in the human EEG and MEG is related to at least two cognitive functions: (1) Numerous studies have shown that binding induces 40 Hz oscillations in humans. (2) The present experiments demonstrate that attention is even more important for gamma activity than binding together the pac-men of a Kanizsa figure. This leads to the question how the two processes of binding and attention interact with each other. Detecting changes in the environment nicely illustrates this interaction.

Figure 14. When pac-men stimuli are bound together a pop-out of the resulting Kanizsa square is perceived. The meaningful Gestalt among randomly arranged pac-men is automatically attended by our visual system serving as an example how binding and attention interact.

Figure 14 shows an example of the binding phenomenon. In visual search displays, which consist of numerous distractor pac-men and one Kanizsa square, the binding of the Kanizsa forming pac-men leads to an automatic pop-out of the Gestalt (Davis & Driver, 1994). That is, the focus of attention is directed towards the locus at which multiple pac-men can be bound to one coherent Kanizsa square. This outcome nicely demonstrates how closely binding and attention are related to each other. In view of this interaction it becomes clear why both processes result in similar electrophysiologic responses. Further research is required to differentiate which responses might be unique to binding on the one hand and attention on the other.

REFERENCES

Adrian, E. (1942). Olfactory reactions in the brain of the hedgehog. *Journal of Physiology* (London), *100*, 459-473.

Başar, E., Başar-Eroglu, C., Karakas, S., & Schürmann, M., (1999). Oscillatory brain theory: A new trend in neuroscience. *IEEE Engineering in Medicine and Biology, 18*, 56-66.

Başar-Eroglu, C., Strüber, D., Kruse, P., Başar E., & Stadler, M., (1996a). Frontal gamma-band enhancement during multistable visual perception. *International Journal of Psychophysiology, 24*, 113-125.

Başar-Eroglu, C., Strüber, D., Schürmann, M., Stadler, M., & Başar, E., (1996b). Gamma-band responses in the brain: A short review of psychophysiological correlates and functional significance. *International Journal of Psychophysiology, 24*, 101-112.

Berger, H. (1929). Über das Elektrenkephalogramm des Menschen. *Archiv für Psychiatrie und Nervenkrankheiten, 87*, 527-570.

Davis, G. & Driver, J. (1994). Parallel detection of Kanizsa subjective figures in the human visual system. *Nature, 371*, 791-793.

Dumermuth, G. (1977). Fundamentals of spectral analysis in electroencephalography. In: A. Remond (Ed.), *EEG informatics. A didactic review of methods and applications of EEG data processing* (pp. 83-105). Amsterdam: Elsevier.

Eckhorn, R., Bauer, R., Jordan, W., Brosch, M., Kruse, W., Munk, M., & Reitboeck, H. (1988). Coherent oscillations: a mechanism of feature linking in the visual cortex? *Biological Cybernetics, 60*, 121-130.

Engel, A., König. P., Kreiter. A., Schillen. T., & Singer. W. (1992). Temporal coding in the visual cortex: New vistas on integration in the nervous system. *Trends in Neurosciences, 15*, 218-226.

Freeman, W. (1975). *Mass action in the nervous system.* New York: Academic Press.

Galambos, R. (1992). A comparison of certain gamma band (40 Hz) brain rhythms in cat and man. In E. Başar & T. Bullock (Eds.), *Induced rhythms in the brain* (pp. 201-216). Boston: Birkhauser.

Galambos, R., Makeig, S., & Talmachoff, P. (1981). A 40 Hz auditory potential recorded from the human scalp. *Proceedings of the National Academy of Sciences USA, 78*, 2643-2647.

Gray, C., König, P., Engel, A., & Singer, W. (1989). Oscillatory response in the cat visual cortex exhibit intercolumnar synchronization which reflects global stimulus properties. *Nature, 338*, 334-337.

Herrmann, C.S., & Mecklinger, A. (2001). Gamma activity in human EEG reflects attentional top-down processing. *Visual Cognition, 8*, 273-285.

Herrmann, C.S., Mecklinger, A., & Pfeiffer, E. (1999). Gamma responses and ERPs in a visual classification task. *Clinical Neurophysiology, 110*, 636 -642.

Jokeit, H., & Makeig, S. (1994). Differing event-related patterns of gamma-band power in brain waves of fast- and slow-reacting subjects. *Proceedings of the National Academy of Sciences USA, 91*, 6339-6343

Jürgens, E., Rösler, F., Henninghausen, E., & Heil, M. (1995). Stimulus-induced gamma oscillations: harmonics of alpha activity. *NeuroReport, 6*, 813-816.

Karakaş, S., & Başar, E. (1998). Early gamma response is sensory in origin: A conclusion based on cross-comparison of results from multiple experimental paradigms. *International Journal of Psychophysiology, 31*, 13-31.

Keil, A., Müller, M.M., Ray, W., Gruber, T., & Elbert, T. (1999). Human gamma band activity and perception of a gestalt. *The Journal of Neuroscience, 19*, 7152-7162.

Miltner, H., Braun, C., Arnold, M., Witte, H., & Taub, E. (1999). Coherence of gamma-band EEG activity as a basis for associative learning. *Nature, 397,* 434-436.

Müller, M.M., Junghöfer, M., Elbert, T., & Rockstroh, B. (1997).Visually induced gamma-band responses to coherent and incoherent motion: A replication study. *NeuroReport, 8,* 2575-2579.

Müller, M.M., Teder-Sälejärvi, W., & Hillyard, S. (1998). The time course of cortical facilitation during cued shifts of spatial attention. *Nature Neuroscience, 1,* 631-634.

Müller, M.M., Gruber ,T., Keil, A., & Elbert, T. (2000). Modulation of induced gamma band activity in the human EEG by attention and visual processing. *International Journal of Psychophysiology, 38,* 283-299.

Squire, L., & Kandel, E. (1999). *Memory: From mind to molecules.* New York: W.H. Freeman.

Tallon, C., Bertrand, O., Bouchet, P., & Pernier, J. (1995). Gamma range activity evoked by coherent visual stimuli in humans. *European Journal of Neuroscience, 7,* 1285-1291.

Tallon-Baudry, C., & Bertrand, O. (1999). Oscillatory gamma activity in humans and its role in object representation. *Trends in Cognitive Sciences, 3,* 151-162.

Tallon-Baudry, C., Bertrand, O., Delpuech, C., & Pernier, J. (1996). Stimulus specificity of phase-located and non-phase-locked 40 Hz visual responses in human. *Journal of Neuroscience, 16,* 4240-4249.

Tallon-Baudry, C., Bertrand, O., Delpuech, C., & Pernier, J. (1997). Oscillatory gamma-band (30-70 Hz) activity induced by a visual search task in humans. *Journal of Neuroscience, 17,* 722-734.

Tiitinen, H., Sinkkonen, J., Reinikainen, K., Alho, K., Lavikainen, J., & Näätänen R. (1993). Selective attention enhances the auditory 40-hz transient response in humans. *Nature, 364,* 59-60.

Index

A

acoustic variance, 65-66
alpha, 133, 153-161, 168, 173
alpha rhythm, 149
anterior cingulate, 25, 34, 91, 101, 118, 138
attention, 1-2, 12-15, 23-36, 41-44, 55-57, 62-63, 65, 69, 75-76, 86-92, 99-109, 117-118, 128-129, 138-139, 142-143, 151-158, 167-181
attention networks, 103, 105
auditory cortex, 10-11, 14, 24-25, 33, 104
automatic process, 43, 63

B

band pass, 45, 154, 175
Berger, 133, 168
beta, 168
binding, 168-172, 179-182
Blood Oxygenation Level Dependent (BOLD), 120-122, 127-128
bottom-up, 62, 167, 172, 178-179

C

central sulcus, 136, 140
change detection, 1-2, 12, 14-15, 28, 35, 61-64, 69-76, 117, 119-120, 129
cognition, 83, 100-101, 103, 105-106, 134, 144
coma, 2, 14-15
complex sounds, 4, 11, 23, 64-65
consonant, 5, 11, 55, 89, 157
covariance, 133, 142-144

D

deblurring, 134
delta, 149, 151, 154, 158-162
delta activity, 158-162
desynchronization, 153-158
deviance effects, 48, 52-55
deviant, 2-14, 23-36, 41-57, 62-76, 123
distractor stimulus, 85-87, 90

E

EEG, 45-46, 88, 100, 119-120, 133-145, 149-154, 157-158, 161, 167-181
electromyography, 174
EMG, 174-175
endogenous, 57
episodic memory, 129, 154-162
ERD, 154
EROS, 24, 34
ERS, 154
event-related potential (ERP), 1-3, 6, 8, 11, 13, 23-36, 41-42, 62, 64-65, 68, 71, 76, 83-86, 89-92, 100-109, 117-122, 126-130, 133-134, 136, 138-139, 142-144, 167, 172-181
Event-Related Potential Covariance (ERPC), 143-144
evoked potential, 46, 48-49, 55, 87-88, 137
exogenous, 2, 42-44, 48, 50, 55-57

F

frontal lobe, 85-86, 90-92, 103, 106, 129, 143
functional magnetic resonance imaging (fMRI), 11, 24, 91, 100-103, 109, 117-118, 120-122, 124, 126, 129, 133-134, 142, 145

G

gamma, 150, 167-181

H

habituation, 109, 167
harmonic, 13, 64, 173-174
hemisphere, 10-13, 108-109, 128, 136-138
hemodynamic, 101, 118, 120-122, 124
hippocampal theta, 152-154
hippocampus, 25, 34, 118, 121, 125, 149-153, 161

I

implicit memory, 160, 162
information processing, 89, 91
inhibition, 14, 99-100

inhibitory, 55, 100, 103-109, 151
intensity, 4, 23, 41, 63, 65-66, 71
intracranial, 2, 15, 24-25, 33-34, 101
inverse, 119-120, 140

K

Kanizsa, 169-170, 178-181

L

language, 9, 11, 13, 15, 102, 122, 136
Laplacian derivation, 134-135
latency, 1-3, 6, 10, 14, 24, 26, 43, 48, 56, 68-70, 85-89, 118-119, 122, 172-173, 177, 179-180
lateral prefrontal cortex, 99-108, 118
lesion, 11, 33-34, 90-91, 100-108, 117, 124, 128, 136
limbic, 99, 101, 105, 161
long-term potentiation (LTP), 152, 160
low resolution electromagnetic tomography (LORETA), 120

M

magnetic resonance imaging (MRI), 91
magnetoencephalography (MEG), 11, 24-25, 33-34, 124, 133, 135, 140, 158, 181
medial temporal lobe, 89-90, 118
memory consolidation, 149, 151, 158-160
memory processing, 88, 161
memory trace, 7-14, 63-64, 161
memory updating, 89, 91-92
mismatch negativity (MMN), 23-35, 41, 62,
multi-stable, 170
music, 67

N

N1, 25-26, 28, 32, 34, 101, 103, 179
N2, 103, 107-109
N4, N400, 122-123, 126-127
neuroimaging, 62, 91, 100-101, 105, 120, 124-125, 129, 133, 136, 144
neurological, 44, 89, 100-103
neuropsychological, 35, 83, 89, 91, 93, 106, 128
novelty detection, 105, 117-118, 125, 129
novelty P3, 104-105, 109, 118-129

O

occipital lobe, 170
oddball task, 84, 87-88, 91-92, 118, 129
orbital prefrontal cortex, 99, 105-108
oscillation, 149-150, 157-158, 167-181

P

P300, 24, 83-92, 102, 118
P3a, 24-34, 62-65, 75-76, 83-87, 90-93, 101-102, 104-105, 109
P3b, 24, 26, 83-93, 102-105, 108, 118-125, 129
parameter, 61, 120, 171
parietal lobe, 91-92, 118, 138
perception, 1, 6, 7, 9, 12-15, 23, 74, 168-170, 179
perceptual, 67, 85, 87, 90, 100, 103, 108, 168-169, 173
PET, 11, 24, 34
phase, 8, 15, 32-33, 152, 156, 168, 171-172, 177-178
phoneme, 4, 9, 11
prefrontal cortex, 24, 99-109, 118, 128-129, 138
psychiatric, 14, 44, 103

R

reaction time, 88, 155, 170, 181
recording, 2, 8, 11, 15, 89-90, 120, 122, 124, 133-136, 140, 142, 144, 173
REM, 149, 158-162
reorienting negativity, 25-29
representational width, 4-5, 12
rhythmic slow activity, 149

S

scalp current density (SCD), 24-25, 28, 34-35
scalp distribution, 85, 90
schema, 87
semantic memory, 155-156, 161
sensory memory, 7, 8, 13, 15, 61, 63-64, 69, 117
sleep, 149-150, 158-160
slow wave sleep (SWS), 149-150, 158
somatosensory, 85, 102, 104, 109, 118, 136, 141, 143, 172
sound representation, 1, 4-7, 12-15
spatial frequency, 45-57, 134
speech, 1, 9, 11-15, 64, 67
supramarginal gyrus, 118, 124, 129, 138

T

target detection, 102-103, 117-118, 130
temporal, 4-8, 13-14, 24-25, 32, 41, 56-57, 66-70, 74-75, 88-92, 99, 101, 105, 108, 109, 117-130, 133-134, 136, 140, 142-144, 171, 177
temporal lobe, 89-91, 118, 126, 160
theta, 138-139, 149-162
three-stimulus paradigm, 84-85
top-down, 75, 167, 178-181
transition frequency (TF), 153

V

visual channel, 55-57
visual cortex, 169
voltage distribution, 24, 26, 29, 33
vowel, 5, 9, 11

W

wavelet analysis, 175
working memory, 87, 91-92, 103, 137-139, 141, 152, 155, 157, 161